MIX
Papier aus verantwortungsvollen Quellen
Paper from responsible sources
FSC® C105338

Dr. Purnima K Sharma
Dr. Dinesh Sharma
Prof. R.K. Singh

Development of Field Propagation Model for Urban Area

Anchor Academic
Publishing

Sharma, Purnima K, Sharma, Dinesh, Singh, R.K.: Development of Field Propagation Model for Urban Area, Hamburg, Anchor Academic Publishing 2017

Buch-ISBN: 978-3-96067-126-8
PDF-eBook-ISBN: 978-3-96067-626-3
Druck/Herstellung: Anchor Academic Publishing, Hamburg, 2017

Bibliografische Information der Deutschen Nationalbibliothek:
Die Deutsche Nationalbibliothek verzeichnet diese Publikation in der Deutschen Nationalbibliografie; detaillierte bibliografische Daten sind im Internet über http://dnb.d-nb.de abrufbar.

Bibliographical Information of the German National Library:
The German National Library lists this publication in the German National Bibliography. Detailed bibliographic data can be found at: http://dnb.d-nb.de

All rights reserved. This publication may not be reproduced, stored in a retrieval system or transmitted, in any form or by any means, electronic, mechanical, photocopying, recording or otherwise, without the prior permission of the publishers.

Das Werk einschließlich aller seiner Teile ist urheberrechtlich geschützt. Jede Verwertung außerhalb der Grenzen des Urheberrechtsgesetzes ist ohne Zustimmung des Verlages unzulässig und strafbar. Dies gilt insbesondere für Vervielfältigungen, Übersetzungen, Mikroverfilmungen und die Einspeicherung und Bearbeitung in elektronischen Systemen.

Die Wiedergabe von Gebrauchsnamen, Handelsnamen, Warenbezeichnungen usw. in diesem Werk berechtigt auch ohne besondere Kennzeichnung nicht zu der Annahme, dass solche Namen im Sinne der Warenzeichen- und Markenschutz-Gesetzgebung als frei zu betrachten wären und daher von jedermann benutzt werden dürften.

Die Informationen in diesem Werk wurden mit Sorgfalt erarbeitet. Dennoch können Fehler nicht vollständig ausgeschlossen werden und die Diplomica Verlag GmbH, die Autoren oder Übersetzer übernehmen keine juristische Verantwortung oder irgendeine Haftung für evtl. verbliebene fehlerhafte Angaben und deren Folgen.

Alle Rechte vorbehalten

© Anchor Academic Publishing, Imprint der Diplomica Verlag GmbH
Hermannstal 119k, 22119 Hamburg
http://www.diplomica-verlag.de, Hamburg 2017
Printed in Germany

Table of Contents

List of Figures	vi
List of Tables	xi
List of Abbreviations	xiii

1. Introduction — 1
  1.1. Historical Overview — 1
  1.2. Cellular Radio Concept — 4
      1.2.1. Frequency Reuse — 6
  1.3. Concept of Handoff — 8
  1.4. Concept of Trunking — 9
  1.5. Statement of Problem — 9
  1.6. Thesis Motivation — 10
  1.7. Literature Review — 10
      1.7.1. Related Work — 15
            1.7.1.1. Field Propagation Path Loss Models — 15
            1.7.1.2. Effect of Climatic Conditions on Radio Communication — 17
            1.7.1.3. Effect of Climatic Condition on Link Budget — 19
  1.8. Contribution of Thesis — 21
  1.9. Outline of Thesis — 22
  1.10. Benefits of Thesis — 23

2. Field Propagation Path Loss Models — 25
  2.1. Background of Field Propagation Models — 25
  2.2. Radiated and Received Power — 26
      2.2.1. Radiated Power — 26
      2.2.2. Radiation Resistance And Received Power — 29
      2.2.3. Friis Transmission Equation — 30
  2.3. Propagation Path Loss — 32
      2.3.1. Causes of Path Loss — 32
  2.4. Mobile Radio Propagation Environment — 33
      2.4.1. Reflection — 34
      2.4.2. Refraction — 34
      2.4.3. Diffraction — 34
      2.4.4. Scattering — 35
  2.5. Field Propagation Path Loss Models — 36
      2.5.1. Indoor Field Propagation Models — 36
      2.5.2. Outdoor Field Propagation Models — 36
            2.5.2.1. Empirical Models — 36
            2.5.2.2. Deterministic Models — 37

2.5.2.3. Stochastic Models	37
2.6. Empirical Model	37
2.6.1. Free Space Path Loss Model	38
2.6.2. Lee Path Loss Model	39
2.6.3. Cost 231 Walfish-Ikegami (W-I) Model	40
2.6.4. Egli Path Loss Model	40
2.6.5. Okumura Model	41
2.6.6. Hata Model	43
2.6.7. Cost 231 Model	43
2.6.8. ECC-33 Path Loss Model	44
2.6.9. Bullington Model	45
2.6.10. Epstein-Peterson Model	45
2.6.11. Stanford University Interim (SUI) Model	45
2.6.12. Walfisch- Bertoni Model	47
2.6.13. Longley Rice Model	47
2.7. Conclusion	48
3. Methodology for Field Data Collection, Analysis & Its Simulation in MATLAB	**49**
3.1. Introduction	49
3.2. Data Collection Tools	50
3.2.1. Nemo	50
3.2.2. Aligent	51
3.2.3. Pioneer	52
3.2.4. X-Tel	52
3.2.5. TEMS	53
3.3. Introduction to TEMS Investigation	54
3.4. GSM Parameters and Their Range	55
3.4.1. Handoff	55
3.4.2. Rx Level	57
3.4.3. Rx Quality	58
3.4.4. Speech Quality Index (SQI)	58
3.5. Requirements for Field Measurement	58
3.5.1. Hardware Requirement	59
3.5.2. Software Requirement	59
3.5.3. Specification of Hardware and Software	60
3.6. Assembling /Installation /Setup Procedure	63
3.6.1. Plugging In Phones and Data Card	63
3.6.2. Plugging In GPS Unit	63
3.6.3. Configuring TEMS Investigation for Data Collection	63
3.6.4. Test Procedure	68
3.7. Various Issues During Measurement	78
3.7.1. Overshooting	78

3.7.2. Bad Quality	78
3.7.3. Bad Coverage	78
3.7.4. Missing Neighbor	78
3.7.5. Dropped Calls	78
3.7.6. Blocked Calls	79
3.7.7. Handover Failure and Delay	79
3.8. Data Analysis and Simulation Tools	79
3.8.1. MapInfo	80
3.8.2. MATLAB	80
3.9. Conclusion	84
4. Performance Analysis of Different Field Propagation Models	**85**
4.1. Introduction	85
4.2. Field Collected/ Measured Data From Drive Test at South Haryana Region	85
4.3. Received Signal Strength and Path Loss in Different Areas	87
4.4. Comparison Between Field Measured Data and Propagation path Loss Models in South Haryana	88
4.5. Comparative Analysis Between Free Space Path Loss Model and Field Measured Path Loss	92
4.6. Comparative Analysis Between W-I Path Loss Model and Field Measured Path Loss	95
4.7. Comparative Analysis Between Lee Path Loss Model and Field Measured Path Loss	100
4.8. Comparative Analysis Between Egli Path Loss Model and Field Measured Path Loss	103
4.9. Comparative Analysis Between Bertoni Path Loss Model and Field Measured Path Loss	105
4.10. Comparative Analysis Between Okumura Path Loss Model and Field Measured Path Loss	109
4.11. Comparative Analysis Betwe en Cost 231 Path Loss Model and Field Measured Path Loss	113
4.12. Comparative Analysis Between Ecc33 Path Loss Model and Field Measured Path Loss	115
4.13. Comparative Analysis Between Sui Path Loss Model and Field Measured Path Loss	120
4.14. Comparative Analysis Between Hata Path Loss Model and Field Measured Path Loss	123
4.15. Conclusion	126
5. The Effect of Climatic Conditions on Field Propagation Model	**127**
5.1. Introduction	127
5.1.1 Related Work	127
5.2. Climatic Conditions of Narnaul (Haryana), India	128
5.2.1 Geographical Location & Climate of Narnaul (Haryana)	129

5.3. Comparison & Field Data Collection During Different Climate Conditions	133
5.4. Development of Propagation Path Loss Model By Considering Different Climatic Conditions	139
5.4.1. Effect of Summer	139
5.4.2. Effect of Winter	140
5.4.3. Effect of Rain	140
5.4.4. Effect of Fog	142
5.5. Comparative Analysis of Field Measured Data, Okumura Model and Developed Okumura Model	143
5.6. Validation of Developed Okumura Path Loss Model	149
5.6.1. By Taking Reference Model	151
5.6.1.1. Fog Attenuation Reference Model	151
5.6.1.2. Rain Attenuation Reference Model	152
5.6.2. By Applying The Developed Model in Another City	156
5.7. Conclusion	161
6. Cell Coverage Area And Effect On Link Budget Due To Climatic Conditions	**163**
6.1. Introduction	163
6.2. Coverage Area	164
6.3. Link Budget and Its Calculations	166
6.3.1. Important Parameters of Link Budget Calculations	169
6.3.1.1 Receiver Sensitivity	169
6.3.1.2 MS Sensitivity	169
6.3.1.3 BTS Sensitivity	169
6.3.1.4 MS & BTS Antenna Gain	170
6.3.1.5 Diversity Gains	170
6.3.1.6 Feeder & Connector Loss	170
6.3.1.7 Pre Amplifier & Booster	170
6.3.1.8 Interference Degradation Margin	171
6.3.1.9 Polarization Loss	171
6.3.2. Uplink Budget And Coverage Area	172
6.3.2.1. Transmitting End	172
6.3.2.2. Receiving End	173
6.3.3. Down Link Budget And Coverage Area	174
6.3.3.1 Transmitting End	174
6.3.3.2. Receiving End	175
6.4. Effect of Climatic Conditions on Link Budget	177
6.4.1. Calculation of Link Budget & Coverage Area in Summer and Winter	177
6.4.2. Calculation of Link Budget and Coverage Area in Heavy Fog (visibility=30m)	179

6.4.3. Calculation of Link Budget and Coverage Area in Heavy Rain (100mm/hr)	181
6.4.4. Calculation of Link Budget and Coverage Area Including All Climatic Effects in Narnaul (Haryana, India).	183
6.5. Conclusion	185
7. Conclusion And Future Work	**187**
7.1. Results & Conclusion	187
7.2. Future Work	190
References	**191**
Appendices	**207**

LIST OF FIGURES

Figure No.	Description	Page No.
1.1	Year Wise Development of Wireless Communication	2
1.2	Global Growths of Mobile and Fixed Subscribers	3
1.3	Illustration Showing the Importance of Accurate Coverage Estimation in Cellular Networks as Compared to Early Land to Mobile System	5
1.4	First Generation Cellular Phone of 1924	6
1.5	Concept of Frequency Reuse	7
1.6	Illustration of Frequency Reuse Concept	8
1.7	Basic of Handoff	8
2.1	The Hertzian Dipole	26
2.2	Voltage Induced at the Receiver Antenna	30
2.3	Illustration of Wireless Communication Showing Path Loss	32
2.4	Phenomenon of Reflection and Refraction	34
2.5	Diffraction in Sharp Edge	35
2.6	Wave is Scattered by a Small Obstacle	35
2.7	Example of Free Space Communication	38
2.8	Median Attenuation Relative to Free Space $A_{mu}(f,d)$ Over a Quasi-smooth Terrain	42
2.9	Correction Factor G_{area} for Different Types of Terrain	42
3.1	User Interface of Nemo Drive Test Tool	50
3.2	E7478A Drive Test System with E6455C IMT2000 Digital Receiver (Agilent Data Collection Tool)	51
3.3	Pioneer Data Collection Tool	52
3.4	Window of XTEL's Data collection Tool	53
3.5	Test Principle Illustrations	59
3.6	TEMS Test Kit used for test drive	60
3.7	TEMS window	64
3.8	Equipment Configuration Windows in TEMS	65
3.9	Equipment Configuration Window	66
3.10	Cell Data Configuring Window	67
3.11	Example of loading of Cell File in Narnaul (South Haryana)	68
3.12	Five basic objects of Map info	81
3.13	Illustration of Map Layer	82
3.14	The Default MATLAB Desktop	83

4.1	Selected Cell Sites for Field Data Collection	86
4.2	Derive Test Result in Cell id NNL001	87
4.3	Variation of Received Signal Strength (dBm) with Distance (Km.) in Three Different Areas of Five Different Cell ids	87
4.4	Variation of Path loss (dB) with Distance (Km.) in Three Different Areas of Five Different Cell ids	88
4.5	Comparison of field measured path loss and Predicted path loss with distance (Site id NNL001)	91
4.6	Comparison of field measured path loss and Predicted path loss with distance (Site id NNL002)	91
4.7	Comparison of Field Measured Path Loss and Predicted Path Loss With Distance (Site id NNL003)	92
4.8	M-file of Free Space Path Loss Model	93
4.9	Comparison Between Fields Measured Path Loss and Free Space Path Loss Model	93
4.10	Variation of Path Loss Between Free Space Path Loss and Practical Field Data for Two Adjacent Cells	94
4.11	Variation of Error Between Field Measured Data and Free Space Path Loss Model	95
4.12	M-file of W-I Path Loss Model	97
4.13	Comparison Between Field Measured Path Loss and W-I Path Loss Model	97
4.14	Variation of Path Loss Between W-I Path Loss and Practical Field Data for Two Adjacent Cells	98
4.15	Variation of Error Between Field Measured Data and W-I Path Loss Model	98
4.16	M-file of Lee path loss model	100
4.17	Comparison Between Field Measured Path Loss and Lee path loss model	100
4.18	Variation of Path Loss Between Lee Path Loss and Practical Field Data for Two Adjacent Cells	101
4.19	Variation of Error Between Field Measured Data and Lee Path Loss Model	101
4.20	M-file of Egli Path Loss Model	103
4.21	Comparison Between Field Measured Path Loss and Egli Path Loss Model	103
4.22	Variation of Path Loss Between Egli Path Loss and Practical Field Data for Two Adjacent Cells	104
4.23	Variation of Error Between Field Measured Data and Egli Path Loss Model	104
4.24	M-file of Bertoni Path Loss Model	105
4.25	Comparison Between Field Measured Path Loss and Bertoni Path Loss Model	107

4.26	Variation of Path Loss Between Bertoni Path Loss and Practical Field Data for Two Adjacent Cells	107
4.27	Variation of Error Between Field Measured Data and Bertoni Model	109
4.28	M-file of Okumura Path Loss Model	110
4.29	Comparison Between Field Measured Path Loss and Okumura Path Loss Model	110
4.30	Variation of Path Loss Between Okumura Path Loss Model and Practical Field Data for Two Adjacent Cells	111
4.31	Variation of Error Between Field Measured Data and Okumura Path Loss Model	111
4.32	M-file of COST 231 path loss model	113
4.33	Comparison Between Field Measured Path Loss and COST 231 Path Loss Model	114
4.34	Variation of Path Loss Between Cost 231 Model and Practical Field Data for Two Adjacent Cells	114
4.35	Variation of Error Between Field Measured Data and Cost 231 Model	115
4.36	M-file of ECC 33 Path Loss Model	117
4.37	Comparison Between Field Measured Path Loss and ECC-33 Path Loss Model	117
4.38	Variation of Path Loss Between ECC-33 Path Loss Model and Practical Field Data for Two Adjacent Cells	118
4.39	Variation of Error Between Field Measured Data and ECC-33 Path Loss Model	118
4.40	M-file of SUI Path Loss Model	120
4.41	Comparison Between Field Measured Path Loss and SUI Path Loss Model	120
4.42	Variation of Path Loss Between SUI Path Loss Model and Practical Field Data for Two Adjacent Cells	121
4.43	Variation of Error Between Field Measured Data and SUI Path Loss Model	121
4.44	M-file of Hata Path Loss Model	123
4.45	Comparison Between Field Measured Path Loss and Hata Path Loss Model	124
4.46	Variation of Path Loss Between Hata Path Loss Model and Practical Field Data for Two Adjacent Cells	124
4.47	Variation of Error Between Field Measured Data and SUI Path Loss Model	126
5.1	Geographical Map of Haryana	129
5.2	Average Rainy Days per Month in the Year 2011-2012	131
5.3	Average Rain Fall in Year 2011-2012	131
5.4	Satellite View & Climate (foggy day) of Narnaul	132

5.5	Average Fog Hours per Day in Year 2011-2012	132
5.6	Variation of Path Loss in Different Climatic Conditions	133
5.7	Error between Measured Data and Okumura Model in Winter	136
5.8	Error between Measured Data and Okumura Model in Summer	136
5.9	Error between Measured Data and Okumura Model in Heavy Fog Climate	137
5.10	Error between Measured Data and Okumura Model in Heavy Rain Climate	137
5.11	Illustration of Collision of Atoms and Molecules	139
5.12	Effect of Sun in Frequency Spectrum	140
5.13	Effect of Rain on Radio Waves	141
5.14	Comparison Between Field Measured Data, Okumura and Developed Okumura Path Loss Model in Winter Climate	143
5.15	Comparison Between Approximated 4th Degree Polynomial Curve of Field Measured data, Okumura and Developed Okumura Path Loss Model in Winter Climate	144
5.16	Comparison Between Field Measured Data, Okumura and Developed Okumura Path Loss Model in Summer Climate	144
5.17	Comparison Between Approximated 4th Degree Polynomial Curve of Field Measured Data, Okumura and Developed Okumura Path Loss Model in Summer Climate	145
5.18	Comparison Between Field Measured Data, Okumura and Developed Okumura Path Loss Model in Heavy Fog Climate	145
5.19	Comparison Between Approximated 4th Degree Polynomial Curve of Field Measured Data, Okumura and Developed Okumura Path Loss Model in Heavy Fog Climate	146
5.20	Comparison Between Field Measured Data, Okumura and Developed Okumura Path Loss Model in Heavy Rain Climate	146
5.21	Comparison Between Approximated 4th Degree Polynomial Curve of Field Measured Data, Okumura and Developed Okumura Path Loss Model in Heavy Rain Climate	147
5.22	Variation of Error Between Field Measured Data, Okumura and Developed Okumura Model in Winter	147
5.23	Variation of Error Between Field Measured Data, Okumura and Developed Okumura Model in Summer	148
5.24	Variation of Error Between Field Measured Data, Okumura and Developed Okumura Model in Foggy Climate	148

5.25	Variation of Error Between Field Measured Data, Okumura and Developed Okumura Model in Winter	149
5.26	Comparison Between Developed Fog Attenuation and Reference Fog Attenuation Model	151
5.27	Difference Between Developed and Reference Fog Attenuation Model	152
5.28	Comparison Between Developed Rain Attenuation and Reference Rain Attenuation Model	153
5.29	Difference Between Developed and Reference Rain Attenuation	153
5.30	Comparison Between Developed Okumura Model and Field Data Taken (Hisar, Haryana, INDIA) in winter season	156
5.31	Comparison Between Developed Okumura Model and Field Data Taken (Hisar, Haryana, INDIA) in summer season	156
5.32	Comparison between Developed Okumura Model and Field Data Taken (Hisar, Haryana, INDIA) in Heavy Fog Condition	157
5.33	Comparison between Developed Okumura Model and Field Data taken (Hisar, Haryana, INDIA) in Heavy Rain Condition	157
5.34	Error between Developed Okumura Model and Field Data taken (Hisar, Haryana, INDIA) in Winter Condition	158
5.35	Error between Developed Okumura Model and Field Data taken (Hisar, Haryana, INDIA) in Summer Season	158
5.36	Error between Developed oOkumura Model and Field Data taken (Hisar, Haryana, INDIA) in Heavy Fog Condition	159
5.37	Comparison between Developed Okumura Model and Field Data taken (Hisar, Haryana, INDIA) in Heavy Rain Condition	159
6.1	Illustration of Link Budget in Mobile Communication	164
6.2	Methodology Used for Prediction of Optimum Coverage Area	166
6.3	Uplink & Downlink Budget	168
6.4	Mast Head Amplifier	171
6.5	Uplink Budget & Flow Chart (Uplink Budget)	173
6.6	Downlink Budget & Flow Chart (Downlink Budget)	174

LIST OF TABLES

Table No.	Description	Page No.
1.1	Evolution of The WLAN Standards	4
2.1	The Parameter Values of Different Terrain for SUI Model	46
3.1	Range of SQI	58
3.2	Some TEMS Supported Mobile Phone's Feature	61
3.3	Signal Strength Measurements at Base Station NNL001 in Month of January (Winter)	70
3.4	Signal Strength Measurements at Base Station NNL011 in Month of May (Summer Temperature: 47^0 C)	71
3.5	Signal Strength Measurements at Base Station NNL001 in Month of July (Heavy Rain)	72
3.6	Signal Strength Measurements at Base Station NNL001 in Month of December (Winter Heavy Fog Condition)	73
3.7	Signal Strength Measurements at Base Station Hisar in Month of January	74
3.8	Signal Strength Measurements at Base Station Hisar in Month of May	75
3.9	Signal Strength Measurements at Base Station Hisar in Month of July	76
3.10	Signal Strength Measurements at Base Station Hisar in Month of December	77
3.11	Main Parts of MATLAB	83
3.12	Different Parts of MATLAB Window	84
4.1	Average Signal Strength Measurements	89
4.2	Average Path Loss Measurements	90
4.3	Error Between Measured and Free Space Path Loss Model	96
4.4	Error Between Measured and W-I Path Loss Model	99
4.5	Error Between Measured and Lee Path Loss Model	102
4.6	Error Between Measured and Egli Path Loss Model	106
4.7	Error Between Measured and Bertoni Path Loss Model	108
4.8	Error Between Measured Okumura Path Loss Model	112
4.9	Error Between Measured and Cost 231 Path Loss Model	116
4.10	Error Between Measured and ECC-33 Path Loss Model	119
4.11	Error Between Measured and SUI Path Loss Model	122
4.12	Error Between Measured and Hata Path Loss Model	125
5.1	Different Seasons of India	128
5.2	Various Climatic Regions of India	129
5.3	Rainfall Statistics for Haryana	130
5.4	Variation of Maximum and Minimum Temperature in Haryana	130
5.5	Average Signal Strength Measurements at Narnaul (Haryana)	134
5.6	Average Path Loss Measurements at Narnaul (Haryana)	135
5.7	Error Between Okumura Path Loss Model and Field Data	138
5.8	Error Between Measured, Okumura and Developed Okumura Model	150

5.9	Error Between Fog Attenuation and Reference Model	154
5.10	Error Between Rain Attenuation and Reference Model	155
5.11	Error Between Measured Data (Hisar, Haryana, India) and Developed Okumura Model	160
6.1	Transmitter Side Specifications (Uplink)	172
6.2	Receiver Side Specifications (Uplink)	173
6.3	Transmitter Side Specifications (Down Link)	175
6.4	Receiver Side Specifications (Down Link)	175
6.5	Coverage Area Calculations in Summer & Winter	179
6.6	Coverage Area in Foggy Days	181
6.7	Coverage Area Calculation in Rainy Days	182
6.8	Coverage Area Calculations by Using Developed Okumura Model	184
7.1	Difference between Measured Field Data to Path Loss Model	188
7.2	Average Error between Currently Measured Data with Okumura and Developed Okumura Model	188
7.3	Coverage Area Calculations Taking Different Parameters	189

LIST OF ABBREVIATIONS

1G	First Generation wireless technology
2G	Second Generation wireless technology
3G	Third Generation wireless technology
1xDV	3G Extension of IS-95B: shared data and voice
1xDO	3G Extension of IS-95B: data only
1xEV	3G Extension of IS-95B: data with circuit-switched voice
1xRTT	3G Extension of IS-95B: one RF channel
ACELP	Adaptive Code Excited Linear Prediction
ADPCM	Adaptive Digital Pulse Code Modulation
AM	Amplitude Modulation
AMPS	Advanced Mobile Phone Service
BCCH	Broadcast Control Channel
BCH	Bose Chaudhuri Hocquenghem *also* Broadcast Channel
BoD	Bandwidth on Demand
BPSK	Binary Phase Shift Keying
BS	Base Station
BTS	Base Transreceiver Station
CC	Convolution Code
CB	Citizens Band
CDMA	Code Division Multiple Access
CEPT	Conference of European Postal and Telecommunications Administrations
COST	Cooperative for Scientific and Technical Research
CT2	Cordless Telephone 2
CTIA	Telephone Industry Association
COST WI	COST Walfisch Ikegami
DCS	Digital Cellular System
DECT	Digital European Cordless Telephone
DQPSK	Differential Quadrature Phase Shift Keying
DS-CDMA	Direct-Sequence Code Division Multiple Access
EDGE	Enhanced Data Rate For GSM Evolution
EIRP	Effective Isotropic Radiated Power
EFR	Enhance Full Rate
ELF	Extremely Low Frequency
ETACS	Extended Total Access Communication System *also* European Total Access Cellular System
ETSI	European Telecommunications Standard Institute
EURO-COST	European Cooperative for Scientific and Technical Research
EHF	Extremely High Frequency
F/TDMA	Hybrid FDMA/TDMA
FDD	Frequency Division Duplex
FDMA	Frequency Division Multiple Accesses
FL	Forward Link
FM	Frequency Modulation
FR	Full Rate

GAN	Global Area Network
GMSK	Gaussian Minimum Shift Keying
GFSK	Gaussian Frequency Shift Keying
GPRS	General Packet Radio Service
GPS	Global Positioning System
GSM	Global System for Mobile Communications
HF	High Frequency
HO	Hand Over
HR	Half Rate
HSCSD	High Speed Circuit Switched Data
HSPDA	High Speed Downlink Packet Access
HSPA	High Speed Packet Access
HSUPA	High Speed Uplink Packet Access
iDEN	Integrated Digital Enhanced Network
IMT	International Mobile Telecommunications
ITU	International Telecommunication Union
ITU-R	ITU's Radio communications sector
IS-54	EIA Interim Standard for U.S. Digital Cellular with Analog Control Channel
IS-95	EIA Interim Standard for U.S. Code Division Multiple Access
IS-136	EIA Interim Standard136 –USDC with Digital Control Channel
ISDN	Integrated Services Digital Network
JTACS	Japanese Total Access Communication System
LF	Low Frequency
LOS	Line of sight
LTE	Long Term Evolution
MSE	Mean square error
MF	Medium Frequency
MS	Mobile Station
NIDS	Network Intrusion Detection System
NMT	Nordic Mobile Telephone
NLOS	Nonlineofsight
NTACS	Narrowband Integrated Services Digital Network
NTT	Nippon Telephone and Telegraph
OVSF	Orthogonal Variable Spreading Factor
PABX	Private Access Business Exchange
PDC	Personal Digital Cellular
PCNs	Personal Communication Networks
PN	Pseudo Noise
PCS	Personal Communication System
PSI-CELP	Pitch Synchronous Innovation CELP
QCELP	Quadrature Code Excited Linear Prediction
QoS	Quality of Service
QPSK	Quadrature Phase Shift Keying
RAN	*Radio Access Network*
RCELP	Residual Code Excited Linear Prediction
RL	Reverse Link

RPE-LTP	Regular Pulse Excited Long Term Prediction
SDCCH	Stand-alone Dedicated Control Channel
SQI	Speech Quality Index
SHF	Super High Frequency
TACS	Total Access Communication System
TCH	Traffic Control Channel
TDD	Time Division Duplex
TETRA	Terrestrial Trunked Radio
UHF	Ultra High Frequency
UMTS	Universal Mobile Telecommunication System
VHF	Very High Frequency
VLF	Very Low Frequency
VSELP	Vector Sum Excited Linear Prediction
WCDMA	Wideband CDMA
WiMAX	Worldwide Interoperability for Microwave Access
WARC	World Allocation Radio Conference

CHAPTER 1

INTRODUCTION

In current era the wireless communication is spreading throughout the world rapidly. The wireless technology has covered each and every area in day to day life. This chapter discusses the historical overview and outline of the thesis along with expected outcome of the research work carried presently.

1.1 HISTORICAL OVERVIEW

Wireless communication is one of the most dynamic and vibrant areas of technology development in the communication field today. To give better understanding, it may be revert from literature of old days that the first outcome of communication started with origin of radio in the year 1680 by Newton's theory of composition of light. According to Newton, light is a composition of various colours and his theory brings the importance of light as a research area of study for many scientists. Later on in 1873, James Clerks Maxwell gave many laws to explain electro magnetism as a result of Poisson's equation using electrostatics, Gauss law equation using magneto statics, Ampere's law equation using electrodynamics, and Faraday law equation using magneto-dynamics. After his research, in the year 1888, Heinrich Rudolf Hertz practically verified the electromagnetism phenomena which Maxwell obtained mathematically [164].

Four years later, in the year 1892, A British scientist Sir William Crookes published a paper on telegraphic communication over long distances using tuned circuits. With the help of Crookes work, Gugliemo Marconi established a radio link over a distance of a small number of miles in 1895. It is the first revolution to the mobile radio industry. The communication with people on the move was made possible by this radio link. Two way radio communication links at frequencies of 30 to 40 MHz were designed from the middle of 1930s [174]. The radio communication gradually increased to include the metric, decimetric and centimetric wavelengths from the year 1930 to 1960 [187]. From the year 1970 frequency modulation was introduced in communication. The analog cellular systems were first developed by Bell Laboratories [186]. In 1979, an effort was made to launch and install first cellular system, i.e. Advanced Mobile Phone Service (AMPS) started at Chicago. Then in 1980 the High Capacity Mobile Telephone System (HCMTS) launched at Tokyo and the Nordic Mobile Telephone (NMT) launched in 1981 at Scandinavia. France's Radiocom 2000 was operational in 1985, similar to United Kingdom's Total Access Communication System (TACS) and Germany's C 450 systems [209]. In the early days of 1990s, low cost cordless system and it got remarkable growth rates. Among these systems cellular played an important role in such growth process, especially after the invention of international digital standards like Global System for Mobile Communications (GSM) and Code division multiple access (CDMA) system (IS-95) [184].

In general, the cellular systems in operation are divided into two categories: the first-generation analog systems and the second generation digital systems. At present, one can observe the quick growth of different types of wireless communication systems for example personal fixed & mobile, and land & satellite. These systems utilize a frequency band from 500 MHz up to 3 to 10GHz. The IMT-2000 third generation cellular mobile system was introduced in 2002.This system relies on cellular techniques and reuses the basic concepts of architecture, functionality and services of these systems [70], [187]. Generation wise the wireless communication is as shown in figure 1.1. The first-generation (1G) mobile systems were analogue, and commissioned in the 1980s. In the 1990s, second-generation (2G) digital mobile systems such as the GSM came in existence. The GSM standard is tremendously triumphant, providing the national as well as international coverage. So, GSM is nowadays the foremost mobile communication system [163].

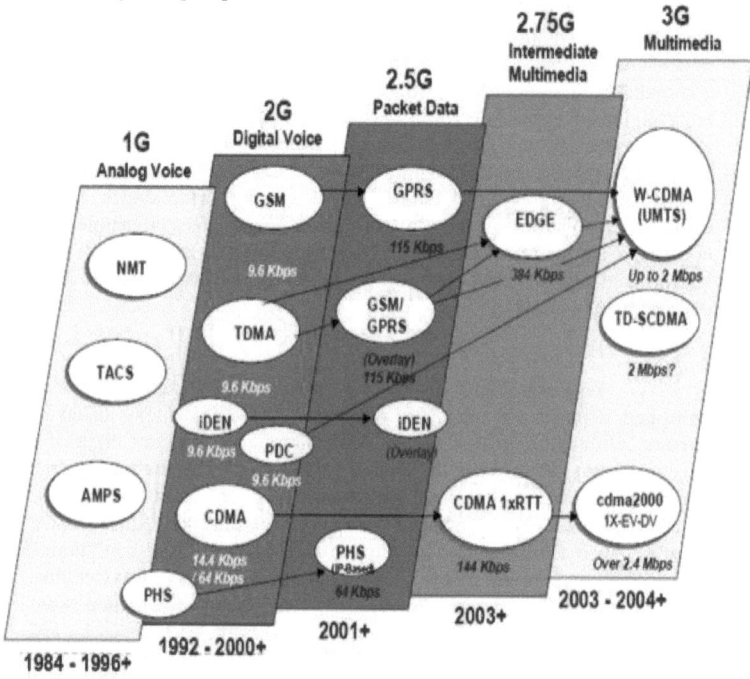

Figure 1.1 Year Wise Development of Wireless Communication

Wireless communication has gained incredible growth in the last few years. The first mobile contributions took place in the early 1980s, and the industry was blooming by 1987. However, the traditional phone technology was analogue. The business take-off by GSM (digital) technology occurred in 1992. In early 1991 hardly one in every thousand people had a mobile phone. But till the end of 2001, approximately 17% people got access of the mobile phone [106]. Within this period the number of countries using a mobile network increased tremendously from 3% to more than 90%. In 2002 the number of mobile subscribers leaves behind the number of fixed-line subscribers. Mobile subscribers outnumbered by 7% fixed line subscribers: Mobile subscribers (million): 1,157 and Fixed lines (million):1,083. Since 2002, the fixed line technology declined, getting closer to the edge of obsolescence. The growth of mobile subscribers is depicted in figure 1.2. It is assumed that this growth will continue to rise, and by 2015 every person will have mobile subscription [163].

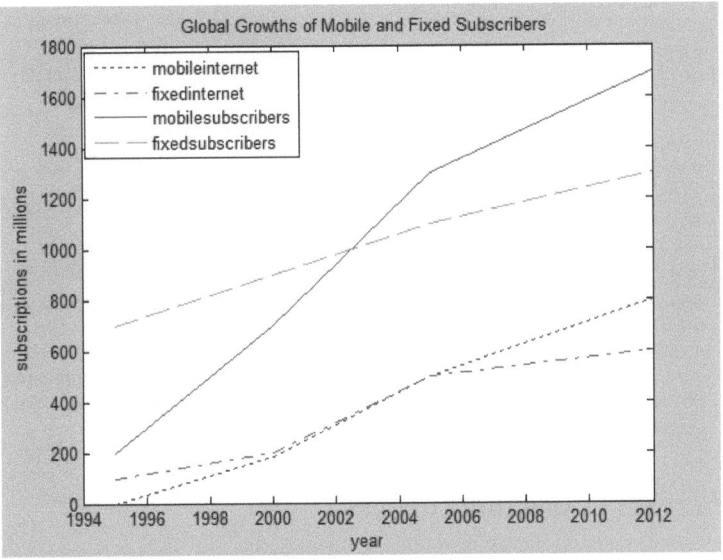

Figure 1.2 Global Growths of Mobile Subscribers

Other than mobile phone communications, Wireless Local Area Networks (WLANs), which came into existence in 1997 only, have also gained tremendous growth. The quick propagation of WLAN hotspots in public places like airport terminal has been amazing. In fact, WLANs have reached into homes, with the help of Digital subscriber line (DSL) and cable access modems resulting in the scenario where number of wireless Internet subscribers will go beyond the number of wired internet users in near future and shown in figure 1.2. The growth of wireless data systems is also seen in many new standards which have recently been developed or are currently under development [163]. Both 1G and 2G systems were intended mainly to offer voice applications, and to support circuit-switched services [167]. However, GSM provides data communication services to users, but the data rates are restricted to only a few tens of kbps. In contrast, WLANs which were designed to offer fixed data network extension in the beginning provide Mbps data transmission rates. The WLAN standard – IEEE 802.11, known as Wi-Fi, was commissioned first time in 1997 and it offered 2 Mbps. Since then the standard has grown numerous times and keeps on increasing as per user requirement for higher bit-rates as shown in Table 1.1. Now days, WLANs can offer up-to 54 Mbps for the IEEE 802.11a/g, and Hiper LAN2 standards operating in the 2.4 GHz and 5 GHz license-free ISM bands. Though, WLANs are not able to provide the kind of mobility, which mobile systems can do [5], [163].

Table 1.1 Evolutions of the WLAN Standards

Year of Establishment	Standard of WLAN Standard	Frequency	Modulation	Bit rate
1997	IEEE 802.11	2.4 GHz	Frequency Hopping and direct spread spectrum	2 Mbps
1998	ETSI Home RF	2.4 GHz	Wideband Frequency Hopping	1.6 Mbps
1999	IEEE 802.11b	2.4 GHz	Direct Sequence Spread spectrum	11 Mbps
1999	IEEE 802.11a	5 GHz	OFDM	54 Mbps
2000	ETSI Hiper LAN2	5 GHz	OFDM Connection oriented	54 Mbps
2003	IEEE 802.11g	2.4 GHz	OFDM Compatible with 802.11a	54 Mbps

Wireless communication should be designed to attain high capacity with limited radio spectrum and it is possible by the Cellular radio concept, which is discussed in the following section.

1.2 CELLULAR RADIO CONCEPT

The concept of cells was introduced in early 1947 by Bell Laboratories in the US; they also gave a detailed proposal for a "High-Capacity Mobile Telephone System" integrating the cellular concept submitted by Bell Laboratories to the FCC in 1971. Still the first AMPS system was set up in Chicago in 1983 [56]. The old system was able to attain a large coverage by means of a simple, high power transmitter in a cell. Base station (BS) was put on the top of mountains or tall towers, so that it could cover a large area. The next Base station BS was put so far away that interference was not a concern. Wireless radio services just in terms of spectrum use alone pretence a much more difficult problem [33]. Severely, it bounds the number of users that could communicate at a time. These were noise-limited systems as numbers of users were limited. The Bell mobile system in New York City in the 1970s was able to communicate a maximum of twelve calls at a time over an area of thousand square miles [125], [186]. The number of calls a mobile wireless system can handle at the same time is essentially determined by the total spectral allocation for that system and the bandwidth needed for transmitting signals used in managing a call. Cellular systems can handle a large number of users over a large geographic area within a limited frequency spectrum. High capacity is attained by using the concept of cell which is a small geographic area and for each cell a single base station is used. Using this concept the same radio channels can be reused by another base station situated some distance away. The entire coverage area can be partitioned into several cells [14]. A cell corresponds to the covering area of single BS transmitter or a small collection of many transmitters. The size of a cell is determined by the transmitter's power.

In this way a single, high power transmitter (large cell) is replaced by many low power transmitters (small cells) which cover only one cell area (a small portion of the service area) as shown in figure 1.3. For mobility a sophisticated switching technique called handoff is used which helps in establishing a call un-interrupted when the user shift, from one cell to another.

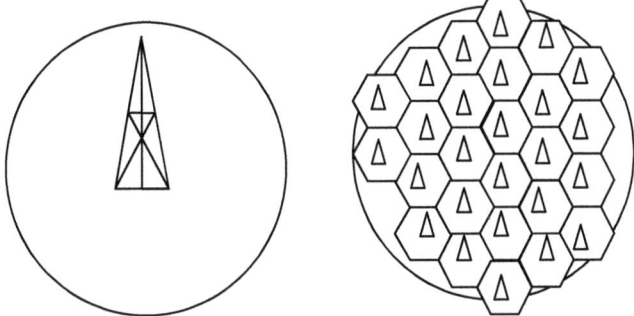

Figure 1.3 Illustrations Showing the Importance of Accurate Coverage Estimation in Cellular Networks as Compared to Early Land to Mobile System

Basic cellular system consists of mobile stations, base stations, and a mobile switching centre (MSC). Mobile switching centre (MSC) is also referred as mobile telephone switching office (MTSO) which manages the activities of the base stations and also connects the entire cellular system to the public switched telephone network (PSTN) [230]. It handles all billing and system maintenance functions. Each communication takes place via radio waves with one of the base stations and for the complete duration of call the mobile station may be handed-off to any number of base stations [231]. Mobile station consists of three units, first one is transceiver, second is an antenna, and third is control circuitry. Among all mobile users in the cell Base stations work as a bridge and helps in connecting the concurrent mobile calls via telephone lines or microwave links to the MSC. It contains a number of transmitters and receivers which concurrently manage full duplex communications. In general it has towers to support numerous transmitting and receiving antennas [232]. Cellular concept also depends on an intelligent allocation and reusability of channels all over a coverage region. These systems are sometimes referred as narrow band systems as these use the concept of frequency reusability. The frequency reuse concept is given in the following section.

Figure1.4. First Generation Cellular Phone of 1924

1.2.1 Frequency Reuse

Cellular notion depends on an intelligent allocation and reusability of channels all over a coverage region. These systems are sometimes referred as narrow band systems as these use the concept of frequency reusability. A group of radio channels are assigned to each cellular base station (BTS) to be utilized within a cell. The design process contains selecting and assigning channel groups to all cellular BTS within a system [80]. Consider a cellular system has a total of S duplex channels available for use. If each cell is allocated a group of k channels, where $k < S$, and if the S channels are divided among N cells into unique and disjoint channel groups which each have the same number of channels, the total number of available radio channels can be expressed as

$$S = k N \qquad (1)$$

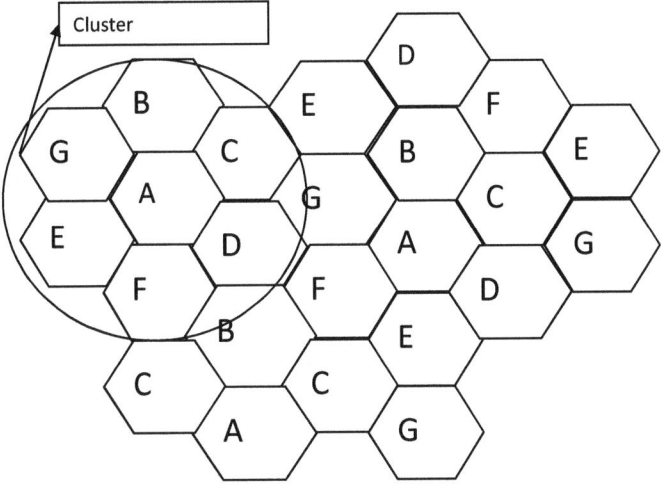

Figure 1.5 Concept of Frequency Reuse

The N cells which collectively use the complete set of available frequencies is called a cluster [90]. If a cluster replicated M times within the system, the total number of duplex channels C can be used as a measure of capacity and is given by

$$C = M k N = M S \qquad (2)$$

The factor N is called cluster size (typically equal to 4, 7, or 12) and it is a function of how much interference a mobile or base station can tolerate while maintaining a sufficient quality of communications [208]. The frequency reuse factor is given by N, since each cell within a cluster is only assigned of the total available channels in the systems. The number of cell, N can be related to the geometry of the hexagons as given below:

$$N = i^2 + ij + j^2 \qquad (3)$$

It means that one has to move i cells along any chain of hexagons and the turn 60° counter-clockwise and move to j cells.

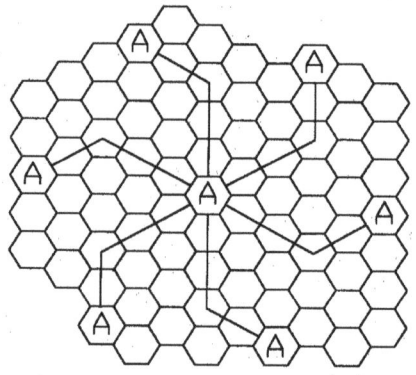

Figure 1.6 Illustration of Frequency Reuse Concept

1.3 CONCEPT OF HANDOFF

When a mobile unit is moving from one cell area to another cell area while a call is in progress, the mobile switching centre (MSC) automatically transfers the call to a new cell area belonging to the new base station [177]. It is defined as the process of changing the current radio channel to a new radio channel [221], [222]. It is a perfect service to all mobile phone consumers while data transfer is in progress. Handoff is an expensive process to execute, so unnecessary handoff should be avoided while ensuring that essential handoffs are finished before a call is terminated due to poor signal level. Handoff includes two major steps; first handoff initiation; In this initiation phase, decision to start the handoff procedure is taken. Second is handoff execution; in this execution phase, a new channel assignment is carried out or if there is no channel available the call is dropped [47], [41]. Basics of handoff are given in the figure 1.6.

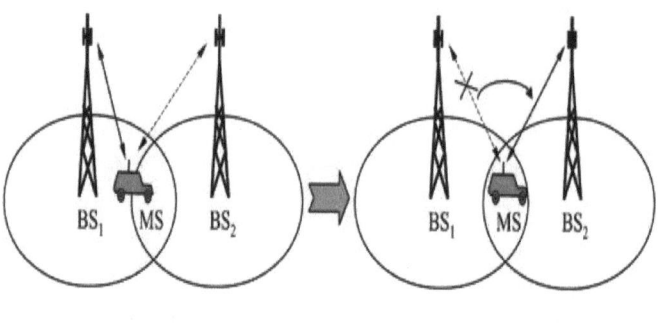

a. Before handoff   b. After handoff

Figure 1.7 Basic of Handoff

1.4 CONCEPT OF TRUNKING

Cellular systems depend on trunking to hold a large number of cellular subscribers in minimum number of channels. The concept of trunking allows a large number of subscribers to share a relatively minimum number of channels by providing access to each user from available channels. In a trunked system, subscriber is assigned a channel on a per call basis, and upon termination of the call, the previously occupied channel is immediately returned to the available channels. The grade of service (GOS) is a measure of the ability of a user to access a trunked system during the busiest hour of call traffic. It is clear that there is a trade-off between the number of available channels and the possibility of a particular user finding that no channels are available during the peak calling time. The number of channels required is determined based the number of subscribers, desired GOS, average call holding time and traffic distribution with time.

Detailed description and problem statement of this thesis is presented in the next section.

1.5 STATEMENT OF PROBLEM

It is investigated and found that severe climatic conditions disturb propagation of electromagnetic signals at higher frequencies [greater than 30MHz] [52]. The disturbance is mainly due to molecular absorption by oxygen for frequencies ranging between 60 and 118 GHz and due to water vapour in 22, 183 and 325-GHz bands [209]. Rain and fog has the most significant impact since the size of the rain drops is of the order of the wavelength of the transmitted signal. It results in energy absorption by the rain drops themselves, and as a secondary effect energy is scattered by the drops. The frequency selective absorption characteristics of the atmosphere can be approximated by a transfer function [95], [131]. In most of practical channels when the signal propagates through the atmosphere affect of many factors on the signal has to be considered along with the free space propagation channel assumption.

Due to those practical channels the incoming radio signal enters the receiver circuitry varies in magnitude. These variations lead to changes in propagation conditions. In acute cases it can even lead to complete cancellation of a signal at the receiving point. These signal variations can occur fast or slow and the speed at which they take place is known as "rate of fading" the reception of microwaves depends on their propagation between a transmitter and a receiver [162]. In the Narnaul city of Haryana (India), the atmosphere is seasonally affected by summer, winter, rain and fog. These different climatic conditions affect the radio wave propagation. The rain and fog, out of these four different climatic conditions plays an important role.

The main objective of this thesis is, therefore, to find out whether, and how, the different climatic conditions are influencing radio wave propagation in GSM band in general and Narnaul, Haryana (India) in particular. To carry out this investigation, the records of radio wave propagation along with path loss during different climatic conditions will be analysed. On the strength of these analyses, a propagation path loss model has been developed by proposing suitable correction factors due different climatic conditions. The validation of this developed path loss model has been verified by taking reference models and by applying practically in different urban area. The effect of these climatic conditions on link budget has been analysed [52].

1.6 THESIS MOTIVATION

The wireless Communication systems: such as radar system, radio navigation, mobile communication, remote sensing, control and recording use radio frequencies as a fundamental transmission medium for their operations.

The need for assessing different climatic conditions such as summer, winter, rain and fog on radio wave propagation is for monitoring of the radio signals and measurements of their field strength and fading characteristics will be analysed. This investigation can also lead the radio planner to a thorough understanding of radio wave propagation in the Narnaul, Haryana (India) for designing mobile communication system. Now, literature review of the thesis will be presented.

1.7 LITERATURE REVIEW

Propagation models are nothing but the combination of analytical and empirical methods. These models are used to calculate electromagnetic field strength for the purpose of wireless network planning during initial establishment. The Harald.T. Friis free space model is mainly used to predict the signal power at the receiver end when transmitter and receiver have line-of-sight condition [186].

The real work on propagation model was initiated in the year 1968 by Okumura. Okumura gave a model to calculate signal strength by collecting measurements surroundings of Tokyo city at frequencies up to 2GHz [168]. The basic source of his model was taken from the free space path loss model. Okumura added the median attenuation (A_{mu}) to the free space path loss in an urban area with a base station height (h_{te}) of 0.2km and a mobile antenna height (h_{re}) of 0.003km. The median attenuation is expressed as a function of frequency (0.1 – 3 GHz) and the distance from the base station (1 -100 km) to the receiver. He also observed and gave modification factors for base station (transmission) antenna $G(h_{te})$ and mobile antenna (reception) $G(h_{re})$. He has obtained some other modification factors in graphical form in suburban and rural area along with urban area [98], [157]. Addition or subtraction of these factors depends on the surroundings and various situations. The entirely empirical nature of the Okumura model shows that the parameters such as frequency, antenna height, range, type of environment, size of the city and the street orientation are restricted to exact ranges determined by the calculated data on which the model is based. This proves that prediction can lead to impractical results if the one or more parameters are used outside the range. Some other constraints also exist with the terrain related parameter.

In the same vein, Hata made some efforts to make the Okumura model easy to apply and establish an experimental mathematical relationship [89] which describes the graphical information given by Okumura. All graphical relations are replaced with mathematical relationships. The problem in Hata's model is its mathematical formulation i.e. it is limited to certain ranges of input parameters. The difference between prediction given by Hata's equation and Okumura curve reveals slight differences that infrequently exceed 1 dB [89]. It was concluded that Hata's method is fairly superior in the urban and suburban areas, but not superior in the rural areas.

William C. Y. Lee gave a model (Lee model) [230] to acquire UHF band propagation characteristics over irregular terrain by use of two approaches: first an area to area algorithm. This approach is based on the equation of straight line presentation of path loss by use of the following parameters: average transmission loss at the range of 1 km, slope of the path loss curve according to plane earth model and an adjustment factor. The standard deviation in predicting the average path loss of this algorithm is 8 dB. Second a point to point algorithm, this approach takes terrain profile into account and it better predicts the variations of the terrain surface. The standard deviation of this algorithm is less than 3 dB.

In the year 1994 European Cooperative for Scientific and Technical research (EURO-COST) has projected COST 231 model to overcome the restrictions in Hata model like frequency range (restricted from 150 MHz to 1500 MHz). In order to accomplish this goal, under COST 231 a huge amount of propagation measurements were performed in the 900 MHz band and 1800MHz band in a large variety of different environments ranging from Pico cells to micro cells and from micro cell to macro cells. Different measurement techniques which are described by J. B. Andersen [25] were used. Much attention was paid to urban micro cellular investigations and to indoor investigations because these cell types will be particularly important for future (Universal Mobile Telecommunications System) UMTS systems.

According to the different objectives explained above, diverse approaches for the classification of measurements were used. Measurements were considered as function of the cell size (e.g. Pico, micro, macro cells), as function of the base station visibility (LOS, NLOS), as function of terrain folds (flat, hilly, etc.), as function of building (large, small) or vegetation density, as function of the mobile speed (from stationary to high speed trains), etc.. Based on this wide-ranging measurement campaigns in European cities, COST 231 has investigated different existing models and has created two new propagation models i.e. COST 231 Hata and COST231 Walfish Ikegami model. These models are suitable for flat terrain and based on the approaches of Walfisch-Bertoni [228], Ikegami and Hata model. The COST 231 Hata model is the extension of Hata model by analyzing Okumura's propagation curves in the upper frequency band and it is limited to macro cell where the base station antenna is on top of the rooftop level of adjacent building.

The COST 231 observed that the evaluation of path loss agrees to a certain extent well with the measurements for base station antenna heights above rooftop level. The mean error is in the range of +3 dB and the standard deviation 4-8 dB [134]. Multi path propagation does not consider in COST-WI .it was observed that the consistency of path loss evaluation decreases if terrain is not flat [135].

A. Ghosh, J. G. et al [76] suggested a way in Broadband wireless access with WiMax/802.16: current performance benchmarks, and future potential. C. F. Ball, et al [27] explained the basic IEEE802.16 (WiMax) 256 sub-carrier OFDM performance for a 3.5 MHz channel in the 3.5 GHz band by link and system level simulations in both interference and coverage limited cellular mobile environment.

Gilhousen K.S, et. Al carried out his research in power-controlled multiple-cell CDMA to increase the cellular capacity [76]. Greg Durgin., and Theodore S. Rappaport., has concluded their results of path loss and building penetration loss measurements in residential areas. Their work determined the effects of shadowing, house construction, and floor plan on the penetration of radio waves into homes [82].

Constantino Perez-Vegay., et al [46] worked on power law path loss model for indoor communications at 1.8 GHZ. In his research he mentioned that the exponent of the distance is treated as a random variable and its behaviour is studied through experiments conducted under a variety of propagation conditions in various buildings. K. Smitha et al [204], presented a modification in ceiling bounce method to find the propagation properties of the channel. Her results clearly show that path loss is a function of separation between the transmitter and receiver.

Aliye "Ozge Kaya et al, also Worked on indoor propagation models and gave a New Path Loss Modeling Approach for wireless Networks inside buildings. The proposed model describes log of diatance will lead to nonlinear-curve-fitting for pathloss [12].

Nagendra Sah. et al [156], surveyed on basic solving techniques behind constraint programming. In particular they concentrated on constraint satisfaction algorithm that are use to solve the constraint satisfaction problem. In his work he focused on various wireless empirical propagation models and solved to find the propagation loss using the constraint satisfaction algorithm.

Comparison of path loss propagation models at 3.5 GHz has been carried out by many researchers in many aspects. V.S. Abhayawardhana et al, worked on this area in Cambridge, UK from September to December 2003 [2]. Josip Milanovic, Rimac-Drlje S, Bejuk K, investigated some empirical propagation models in different terrains as function of antenna height parameters [118]. Basharat worked on CDMA versus IDMA subscriber cell density and explained that beyond third generation (B3G) and fourth generation (4G) communication systems require bandwidth efficiency and low complexity receivers to accommodate high data rate and large number of users per cell. He provided the recommendations for why interleave division multiple access (IDMA) stands out among all the present day multiple access systems [28].

Ubom, E.A., Idigo, V. E., Azubogu, A.C.O., Ohaneme, C.O., and Alumona, published a paper in which he has suggested statistical path loss models derived from experimental data collected in Port Harcourt in South-South region of Nigeria from 10 existing microcells operating at 876 MHz. The results of the measurements were used to develop path loss models for the urban (Category A) and the suburban (Category B) areas of Port Harcourt [220]. M. A. Alim, M. M. Rahman, M. M. Hossain, A. Al-Nahid explained in their paper that Channel properties influence the development of wireless communication systems. They also explained that in mobile radio systems, path loss models are necessary for proper planning, interference estimations, frequency assignments and cell parameters in a wireless system [11]. Purno Mohon Ghosh, Md. Anwar Hossain, A.F.M. Zainul Abadin, Kallol Krishna Karmakar Many Path loss models for macro cells like Hata Okumura, Walfisch-Ikegami and Lee. The received signal strength was calculated with respect to distance and model that can be adopted to minimize the number of Handoffs [75].

Mohammed Alshami, et al, analyzed and compared the path loss values and determined the link budget, power outage probability and WiMAX cell coverage area. His research work discusses and implements WZ Okumura, Hata, Cost- 231, Ericsson, Erceg, Walfish, Ecc-33, Lee and the simplified free space path loss models. All the models applied in his paper are used to predict the propagation loss at WiMAX cell-edge [153]. According to ECC Report 33, [52] the analysis of the existence of FWA cells is in 3.4–3.8 GHz frequency band. The level of this Report give procedure for efficient, technology independent operation of 3.5 GHz (or 3.7 GHz) Point-to-Multipoint (PMP) Fixed Wireless Systems (FWS) [62].

Julio C. Costa, prepared his thesis to share some insight on the propagation characteristic of the radio path in the Tampa Bay area. modified models in the Tampa Bay area were presented, including a specific modified model to support bridges in the Tampa Bay area. These models will help more accurately predict coverage and interference within the area [120]. Pu wang., et al [180] explained about sand storm effects on signal strength of wireless communication signals in their signals. Many ideas from "S. Dey and J. Evans, about Optimal power control over multiple time-scale fading channels with service outage constraints is given for parallel fading channels with fast Rayleigh fading, as a function of the slow fading gains [54].

Tapan K Sarkar., et al [211], made a survey on various propagation models that can provide good estimates for fading channels. Adegoke et al. (2008) did an evaluation of the performance of GSM operators using Nigeria as a case study and examined the problems in front of the industry. Their main focus devoted on improving the performance of the network elements [4]. Adebayo T.L. et al. investigated the propagation path loss characteristics of GSM signals in Benin city, Nigeria using 15 different environments. Later the data was analyzed to calculate path exponent [3]. Gorazd kandus et al. presented a paper on pathloss analyses in tunnels and under ground corridors and it is very useful in pathloss analysis [81]. Andreas F. Molisch gave definition of pathloss as the average attenuation (reduction in power) of a radio signal as it propagates and includes the propagation losses caused by free space and effects due to absorption, diffraction, and others [18].

E.Reusens., et al [189], developed Path loss models for the on body channels. J. De Bruyne, et al [57] investigated the actual measured performance of an 802.16-based system. A measurement methodology for evaluating the performance is proposed, which is then used for studying and comparing the results of different scenarios. More exclusively, the influences of varying the modem height from 2.5 m to 6 m and base station height from 15 m to 45 m are analysed and discussed in this paper, and it will be shown that only the latter one has a significant effect on the coverage and the performance. Finally, as the system supports link adaptive modulation and coding, the results of its effectiveness are discussed.

Liao D. and K. Sarabandi., [130] calculated the far-field radiation from an infinitesimal electric dipole embedded inside a truncated vegetation layer above a dielectric ground plane. Lorne C. Liechty [133] carried out an experiment at campus of the Georgia Institute of Technology. Using the measurements in that area, they observed that a simplistic direct-ray, single path loss exponent, adaptation of the Seidel -Rappaport model can yield satisfactory results in terms of accuracy of model for outdoor microcell environments. Armoogum V., et al [19], made a qualified Study of Path Loss using active Models for Digital Television Broadcasting for Summer Season in the North of Mauritius and the Results showed that the path loss is not constant at various locations for a constant distance around the base station.

Hazer Inaltekin., et al [91] , explained the effects of the singularity in unbounded path-loss models on network performance. Meng, Y. S. et al [147], [148], [149], worked in forest environment and explained radio wave attenuation in those environments. His report gives information about the physical processes when the radio signals propagating through a deep forest. Tan I., et al. presented a GPS-enabled channel sounding platform for measuring vehicle-to-roadside wireless channels. This platform was used to conduct an extensive field measurement campaign involving vehicular wireless channels across a wide variety of speeds and line-of-sight conditions [210].

Turkan ERBAY DALKILIC., et al [218] completely worked on Fuzzy adaptive neural network approach to path loss prediction in urban areas at GSM frequency (900 MHz) band. Lkhagvatseren. T., and Hruska. F., [132], explained propagation of RF signal from frequency 1 to 8 GHz range. Noman Shabbir., et al , published about the radio propagation models used for the upcoming 4th Generation (4G) of cellular networks. The radio wave propagation model or path loss model plays a very vital role in planning of any wireless communication system. In his paper, a comparison is made between different proposed radio propagation models [166]. Zhi ren., et al, studied the effects of Rayleigh fading, path loss, and shadowing fading on wireless mobile networks and implement ed modelling and simulation of Rayleigh fading and shadowing fading with OPENT in their paper [238]. The literature survey on related work of this thesis is presented in the following section.

1.7.1 Related Work

1.7.1.1 Field Propagation Path Loss Models

In 2008 Faihan D. Alotaibi and Adel A. Ali [15], [16], [17] has introduced modification of the Lee path loss empirical model using an automatic LS algorithm. Their calibration is based on conventional signal measurements taken in Riyadh, Saudi Arabia on TETRA network. To verify the LS algorithm method and for reasonable performance estimation they have compared the measured signals against predicted ones using the tuned model in addition to the three most widely used empirical path loss models these are Hata, ITU-R and COST-WI NLOS. They found that performance of the tuned Lee model is the best, as root mean square error is the lowest compared to the other mentioned models. Also they have found that Hata, ITU-R and COST-WI NLOS empirical models expect too much of the path loss for both urban and suburban environments. It is obviously clear from their studies that the tuned Lee model shows the closest agreement with the measured results, while ITU-R and COST-WI NLOS models perform better than the Hata empirical model.

A semi empirical propagation model for the frequency band from 850 MHz to 900 MHz has been proposed by Juan M. Casaravilla and Gabriel A. Dutra in 2009. To evaluate the performance of the MOPEM model against the COST-WI model, Juan has made a close comparison between measurements of MOPEM model and COST-WI model. The outcome reveals that proposed model i.e. MOPEM is more precise than COST-WI model for the region. The author also recommended significant modifications for COST-WI model with regard to the reference model i.e. MOPEM model [38]. Vinko Erceg., et a has presented a statistical path loss model for 1.9-GHz wireless systems in suburban environments. The path loss it predicts can be either the local mean (time-averaged) value for a mobile system or the broadband value for a fixed system. The model makes distinctions among different terrain categories. The result is a general statistical framework for describing path loss that can be upgraded with further measurements [224], [225].

Zia Nadir [157], [158], [159] has investigated Hata model for GSM band in Oman. For survey he conducted the measurement in Salalah, Oman with the help of TEMS tool. After determining the path loss of the practical measurements for each distance, the study was carried by him, in order to make a comparison between the measurement and Hata model. The comparison results clearly explain that the measured path loss is less than the predicted path loss by a difference varying from 4 to 20 dB. Then, mean square error (MSE) was calculated between measured path loss value and those predicted by Hata model. The mean square error (MSE) was found 112.459dB but the acceptable range is up to 6 dB. Zia Nadir investigation shows that Hata propagation model may not be completely adapted in Oman so improvement of Hata model in the open area has been recommended.

In 2010, R. Mardeni and K. F. Kwan [142], [143], [144] analysed the Hata model, Egli model and COST-WI model. They have taken the outdoor measurements in Cyberjaya, Malaysia in order to make a path loss comparison with these existing models. The frequency ranges used for measurement are from 400MHz to 1800 MHz, covering CDMA [121], GSM900 and GSM1800 technologies. From the comparison they found that the performance of the Hata model is the superior as compared to other mentioned models. Generally, COST-WI and Egli model present better than Hata for suburban area in Malaysia. Roelens, in 2005 worked on pathloss models for wireless narrow band communication near biological tissue [191].

Shoewu, O and Adedipe, A. [200], [201] shows that the Hata model for radio wave propagation is very efficient for radio wave propagation path loss prediction in suburban areas in Northern part of Nigeria. They have used a GSM base station operating at 900MHz band for the experiment in a typical suburban area within the Northern part of Nigeria. The field measurement results were compared with Hata model for rural and suburban area. The outcome obtained by them point out the slightest variation with Hata model for suburban areas.

In 2010, Mardeni, R and Lee Yih [141] investigated the Free space model, Okumura model, Egli model and Hata model for urban outdoor coverage in Kuala Lumpur, Malaysia for Code Division Multiple Access (CDMA) system. Mardeni, R and Lee Yih conducted a measurement test in Kuala Lumpur for CDMA system. Subsequent to the contrast between above model and test result, they have found that Okumura model is the preeminent for CDMA in the region. In [239], Zhu and McNair presented cost functions that account for the dynamic values that are inherent to vertical handoff and incorporate a network elimination factor to potentially reduce delay and processing power in the handoff calculation.

1.7.1.2 Effect of Climatic Conditions on Radio Communication

Abdullahi, mainly concentrated on the wireless network features and their effects on radio propagation quality [1]. Olagoke worked on traffic of NITEL GSM network. Main features of his work are network Optimization, network monitoring and network handling by continuous measuring, and analysing the traffic data [169]. In the year 2001, Isaac I. Kim, Bruce McArthur, and Eric Korevaar, worked on laser beam propagation at 785nm and 1550nm in fog and haze for optical wireless communication and found 785nm, 850nm and 1550 light suffer from atmospheric attenuation. Their observation of wavelength, attenuation in fog is important, because fog, heavy snow, and extreme rain are the only types of climates that can disturb communication links [101].

J. A. Weinman, R. Davies and R. Wu, concluded that Water or ice particles blown from the ground into the atmosphere take the form of liquid water as in rain, and fog as in clouds. electromagnetic waves travelling through air containing precipitation are scattered and absorbed by the particles of ice, snow or water. Water has larger dielectric constant and it scatters electromagnetic wave more strongly than ice [229,234]. In addition to above conclusion Akira ishimaru in 1978, gave a conclusion that dielectric loss and the attenuation due to thermal dissipation is greater for water particles than for ice particles. The conclusion given by Akira ishimaru is again discussed in 2004 by Jonathan H. Jiang and Dong L. Wu [103], [113].

David M. Pozar, in his book mentioned that, attenuation is caused by the absorption of microwave energy by tropospheric gases when the frequency coincides with one of the molecular resonances of water or oxygen in the atmosphere [179]. Transmission of microwave signals above 10 GHz is vulnerable to precipitation, as has been shown by many researchers over several decades [43], [65]. In his work he considered radio path where the Fresnel zone is partially filled with rain droplets. Each particular raindrop will contribute to the attenuation of the required signal. The actual amount of attenuation is dependent on the frequency of the signal and the size of the raindrop. The two main causes of attenuation are scattering and absorption. When the wavelength is fairy large relative to the size of raindrop, scattering is predominant. Conversely, when the wavelength is small compared to the raindrop size, attenuation due to absorption is dominant [110].

Frey in his paper explained that Propagation of radio waves above 10 GHz through the atmosphere is greatly influenced by effects of molecular resonance and precipitation [212]. Lakshmi Sutha Kumar, Yee Hui Lee, and Jin Teong Ong concluded that lower rain rates, does not affect the communication links [129]. Due to the applied high carrier frequency (above 20 GHz) besides the existing interference and noise the main degrading factor in these systems is attenuation caused by precipitation, especially rain attenuation [104]. Dougherty H.T. et al, includes attenuation due to both rain and gases. Dutton has developed an updated computer program to predict the rain attenuation, cloud attenuation and attenuation due to atmospheric gases [58]. The rice and Holmberg gave model which is based upon two rainfall types: convective ("Mode 1", thunderstorm) rains and stratiform ("Mode 2", uniform) rains [233].

Crane and Blood gave a Prediction Model which includes path averaging implicitly, and adjusts the isotherm heights for various percentages of time to account for the types of rain structures which dominate the cumulative statistics for the respective percentages of time. Both forms will be described here because the latter is the recommended form for use by system designers, but the earlier form is computationally easy to implement and allows rapid computation with a hand-held calculator [50]. In 2009 Shkelzen Cakaj analyzed the rain attenuation impact on the performance of the respective ground station [36]. G. H. Bryant, I. Adimula, C. Riva and G. Brussaard described a new model for determining rain attenuation on satellite links and explained Rain attenuation statistics from rain cell diameters and heights [35].

Asoka Dissanayake et al, results indicate that the rain attenuation element of their model provides the best average accuracy internationally between frequencies 10GHz and 30 GHz [22]. Oyesola Olayinka Olusola in his thesis gave a case study on mobile radio wave propagation and Shittu (2006) carried his research to cellular mobile radio propagation characteristics within urban and rural environments. He explained the effect of various propagation losses on GSM signals [170], [198]. Bruce (2006) did research on the prediction of seasonal effects on cellular systems in the United States. His work mainly focussed on the effect of foliage in wet seasons on radio wave propagation [34]. Helhel et al. considered the effect of dry and wet seasons on GSM signals. The research conducted in Turkey inside a forested area. The losses observed in his experiment were foliage losses. These losses (Dry and Wet seasons) were compared and gave a conclusion that losses are greater during the wet season [92].

Fraizer explained the effects of foliage on the propagating signal in his research [68]. Mohammed et al. (2006) took many measurements for signal attenuation through Date Palm trees in North Abu Dhabi, United Arab Emirates. His measurements in this region showed significant additional losses of up to 20 dB due to foliage [152]. Douglas (1973) did his research during two different seasons in a suburban area of the United States of America found that the presence of increased foliage reduced the received signal strength to typical values of about 10 dB and he stated that foliage is a significant feature which affects propagation in suburban and rural areas but can be neglected in most urban areas [59].

Batariere M.D. et al. continued the similar work and carried out measurements of path loss due to tree foliage for a carrier frequency of 3.676 GHz in outside Chicago, Illinois. He found lots of difference in path loss calculations in between Urban and rural areas, Because rural area consists more foliage than urban areas and in the other case Urban areas have more high-rise buildings. These differences cause losses in radio propagation characteristics between urban and rural areas [29]. Bruce (2006) stated that the effect of foliage losses would be higher in rural areas comparison with urban areas. Bruce's studies however did not focus on this difference

[34]. Vaclav Kvicera and Martin Grabner explained that the influence of climatic conditions effect the electromagnetic signals all the way through the medium of the lower troposphere. He gave many suggestions for efficient planning and utilization [127]. The aim of Shkelzen Cakaj is to analyse the rain attenuation impact on the performance of the respective earth station. Rain attenuation depends on geological location where the satellite ground station is implemented [36].

M. Sridhar et al mentioned various impairments which causes the signal fade in his technical paper and also explained that rain attenuation is dominant in those impairments. He used ITU-R model to predict the rainfall rate and attenuation due to rain [206]. Dong you choi presented the results of measurements of rain-induced attenuation in vertically polarized signals propagating at 12.25 GHz for the duration of definite rain events [wet season of 2001 and 2007 at Yong-in, Korea]. The rain attenuation over the link measured practically and compared with loss/attenuation got by the ITU-R model [101]. Kiran Ahuja and Manoj Kumar explained the importance of empirical & physical propagation models to calculate the path loss by predicting the received signal strength at a specific point in space by considering the particular propagation surroundings. They considered Free Space, FCC, ITUR370, Okumura (HATA) and physical models viz. TIREM, Anderson 2D v1.00 in their experiment [8].

1.7.1.3 Effect of Climatic Conditions on Link Budget

T.-S. Chu, and Larry J. Greenstein, carried out his research in link budget analysis and explained that the median propagation loss in the personal communication services (PCS's) band is greater than in the cellular band. They carefully examined all factors involved to quantify the link budget differences between two bands in three different terrains (urban, suburban, and rural.) and found that link budget parity can be accomplished in all three environments with quite reserved remedies. This remedy includes the use of tower-top electronics and minor increases in downlink power [45]. Peter J. Black and Qiang presented his paper on link budget of cdma 2000 1xev-do wireless internet access system. It contains the analysis and simulation results for a 1xEV-DO link budget and also contains the traditional fixed rate CDMA link budget calculation including link adaptation and multi-user diversity gains. The main conclusion of his paper is that 1xEV-DO provides a link budget advantage over IS-95-A link [32].

K.Ayyappan, P. Dananjayan prepared his paper to share some insight on the propagation characteristic of the radio path in Pondicherry, INDIA and gave a conclusion that Radio propagation is necessary for up- coming technologies with proper design, operation and management strategies for several wireless networks. Exact description of radio channel through key parameters and a mathematical model is significant for predicting signal coverage. Path loss models such as Hata Okumura, COST 231 and ECC 33 models are analyzed and compared their parameters. This paper proposed a path loss model for a highway between Pondicherry – Villupuram. Analysis also included the link budget calculation [24]. In the same vein P.Saveeda,E.Vinothini,Vardhi Swathi and K.Ayyappan continued this research and concluded that Radio propagation profoundly site specific and varies considerably based on many parameters. Some other constraints also exist with the terrain related parameter. They analyzed and compared the path loss values and determined the link budget [194].

Dr. S. A. Mawjoud studied about the parameters which affects the communication performance of a wireless channel. He took a basic cell, by estimating the affecting parameters on the signal power level in the uplink and downlink at all practical circumstances considering the factors causing fading and other losses in the signal power [145]. Dr. Joe Montana, Dr. James W. LaPean and Dr. Jeremy Allnutt explained link budget in their way along with the effects of atmosphere on link budeget [114].

Joshua D. Griffin and Gregory D. Durgin worked on Complete Link Budgets for Backscatter-Radio and RFID Systems. They explained that communication by modulating signals scattered from a transponder (RF tag) - is basically different from conventional radio. Conventional radio involves two distinct links: the power-up link for powering passive RF tags, and the backscatter link for describing backscatter communication. Their article presented four link budgets which can give explanation for the major propagation mechanisms of the backscatter channel. they demonstrated the Use of the link budgets by a practical UHF RFID. The benefits of future 5.8 GHz multi-antenna backscatter-radio systems are shown [83].

LUIGI MORENO explained fundamental parameters useful to describe radio site installations in his book "POINT-TO-POINT RADIO LINK ENGINEERING". The whole book covers the main topics in Radio Propagation and Point-to-Point radio link engineering [136].

Sebastian Büttrich, concluded in his work that A good link budget is the basic requirement of a well functioning of a radio propagation link. He mentioned many Losses takes place in every element along the radio wave signal transmission path [195]. Tranzeo wireless technologies provide the reader with an overview on how a link budget is calculated. They also explained that Available and permitted output power, available bandwidth, receiver sensitivity, antenna gains, radio technology, and environmental conditions are some of the major factors that may impact system performance. For large scale network operations, detailed site survey and network designs are highly recommended like link budget analysis [216].

The coverage area, link budget and its calculations are discussed in the following section.

1.8 CONTRIBUTION OF THESIS

By combining analytical and empirical methods the field propagation models is derived. The field Propagation models are used for calculation of electromagnetic field strength for the purpose of wireless network planning during preliminary deployment. It describes the signal attenuation from transmitter to receiver antenna as a function of distance, carrier frequency, antenna heights and other significant parameters like terrain profile (e.g. urban, suburban and rural).

The aim of the present research is to analyse field propagation models like Hata, Okumura, SUI, Egli, Cost 231, ECC-33, LEE, W-I and Free space field propagation path loss models for efficient network planning and selection of best fit model for prediction of exact path loss in the area of NARNAUL (Rajasthan, India). Further the different climate as summer, winter, fog and rain conditions has been analysed and attenuation by these climatic conditions be added. At the last the effect of these climatic conditions has been taken in link budget. The outline of contribution is as follows:

- The field propagation path loss models available in the open literature have been reviewed.
- Field data of BSNL GSM network has been collected over the period of two years using TEMS navigation tool in different climatic conditions of Narnaul, Hisar (Haryana, India).
- The analysis of Hata, Okumura, COST 231, ECC-33, SUI, Free space, Lee, W-I and Egli path loss models has been carried out using MATLAB.
- The selection of best fit field propagation path loss prediction model for Narnaul (Haryana, India) region has been estimated on the basis of the path loss and handoff process.
- After analysing the results it has been found that the Okumura path loss model gives better results as compared to other field propagation path loss prediction models in Narnaul (Haryana).
- The attenuation due to the different climatic conditions has been analysed.
- On the basis of performance analysis a suitable development has been done in Okumura field propagation path loss model by considering the area factor and climate factor.
- The developed Okumura model is proposed for Narnaul (Haryana) in all climatic conditions.
- The field measured data has been compared with the developed Okumura path loss model in different climatic conditions.
- For the validation of the developed Okumura model, a comparative study of different attenuation has been done with models of Rain attenuation and Fog attenuation which are given by M. Sridhar et.al. and Altshuler.
- For practical validation of the developed Okumura model, a comparative analysis has been done in Hisar (Haryana, India).
- The effect of climatic attenuation on link budget has been analysed and estimate the coverage area.

Based on the results of current research GSM network planner can adopt field propagation model after analysing their performance. The proper selection of propagation path loss model provides better coverage area and quality to the customers [178]. As the climatic conditions also affects on the radio wave propagation. So developed field propagation path loss model has been implemented to provide more precise network planning [175].

1.9 OUTLINE OF THESIS

Network planning is a complex process consisting of several steps and involves through estimation by field propagation models. The purpose of the network design is the extension of existing network or it has been used for establishment of new network. The basic requirements for network planning are to meet coverage area and quality. The environmental factors such as area, foliage, rain, fog etc also greatly affect network planning. Careful network planning has become important with the rising development, coverage and congestion of wireless area networks. Many researchers made effort to analyse the performance of field propagation path loss models for better network planning. A field propagation model has been estimated for performance analysis in terms of path loss and handoff. In this thesis analysis has been made to achieve prediction accuracy of different field propagation path loss models for Narnaul (Haryana, India). On the basis of this analysis a best fit field propagation model has been selected and developed on the basis of different climatic conditions. The effect of different climatic conditions has been analysed and a change in link budget has been proposed for coverage area.

Chapter 2 discusses radio propagation principles and propagation mechanisms, primarily for modelling any radio channel in outdoor wireless communications along with outdoor field propagation path loss models which works on GSM frequency band.

The chapter 3 describes measurement procedure, hardware, methodology and logistics of all field propagation campaigns which were conducted. Now a day's many data collection tools are available in the market in which TEMS data collection tool has been opted for the for the present research work because it offers many advantages as compared to other data collection tools. The MATLAB and Mapinfo are used after the field data collection. The measurement procedure is based on the use of a drive test system. A drive test simply means drive and test while roaming in the wireless network in a car. The drive test system provides the insight to the performance of a network particularly in terms of RF coverage.

In chapter 4, the comparative analysis of different field propagation path loss models has been carried out with field measured data during drive test in urban area. On the basis of the analysis it has been conclude that the Okumura path model is the best fit path loss model for Narnaul (Haryana, India).

As the climatic conditions plays very important role in radio wave propagation. So the climatic attenuations specially rain and fog attenuation has been taken into account and developed a field propagation model. From the analysis which carried out in chapter 4, it was observed that the Okumura path loss model is best fit propagation model for Narnaul (Haryana, India). But still there is some significant error between measured field data in different climatic conditions and the predicted values by Okumura model. So to develop more precise model, the effects of Fog and Rain attenuation has been discussed in chapter 5.

In chapter 6, the cell coverage area and effect on link budget due to different climatic conditions has been discussed. Further the down link and up link budget calculations has been done.

The thesis ends with Chapter 7, where conclusions are drawn, and recommendations made, about the performance of field propagation path loss model and its suitability in Narnaul (Haryana, India)) on GSM frequency band along with coverage area and calculations of link budget.

1.10 BENEFITS OF THESIS

The field propagation path loss models are designed for a particular area or terrain. For example the Okumura model for Urban Areas is a Radio propagation model that was formulated using the data collected in the city of Tokyo (Japan). The model is ideal for using in cities with many urban structures but not many big blocking structures. So when these field propagation path loss models are used in environment other than where they has been designed, these field propagation models do not predict good results. The network establishment is very expensive process. If these field propagation models are used for network planning in any other place where these models are actually designed, than the correction factor is needed for that particular area.

This study makes a recommendation on necessary mobile communication system design. Based on the results of current research GSM network planner can adopt field propagation model after analysing their network performance. The proper selection of propagation path loss model provides better coverage area and quality to the users. As the climatic conditions also affects on the radio wave propagation. So developed field propagation path loss model can implemented to provide more precise network planning and calculations of Link budget.

CHAPTER 2

FIELD PROPAGATION PATH LOSS MODELS

The theoretical origins of the propagation phenomena and the received signal power conception are described in this chapter. The basic field propagation path loss models have been described here.

2.1 BACKGROUND OF FIELD PROPAGATION MODELS

For calculating electromagnetic field strength, different propagation models have been used for wireless communication network planning during initial exploitation. The field propagation path loss models are derived by combining analytical and empirical methods. These models describe the signal attenuation as a function of distance (between transmitter and receiver), carrier frequency of the signal, height of antennas and other important parameters like topography profile.

In line-of-sight (LOS) condition, the models such as Harald.T.Friis and free space model are used to predict the signal power at the receiver end. For initial coverage deployment, the conventional Okumura model is used in urban, suburban and rural areas for the frequency range from 200 MHz to 1920 MHz. Hata-Okumura model which is known as Hata model, a developed version of Okumura model, also extensively used for the frequency range 150 MHz to 2000 MHz in a build up area.

Comparison of path loss models at different frequencies has been analysed by many researchers in various respects. In Cambridge (UK) [2], the fixed wireless access (FWA) network researchers investigated several empirical propagation models in different terrains as function of antenna height parameters. Another measurement was taken by considering line of sight (LOS) and at NLOS conditions at Osijek in Croatia in the year 2007 [118]. Coverage and throughput prediction were considered with respect to modulation techniques in Belgium [117].

For designing propagation model, different parameters like radiated power, radiation resistance, received power etc., are very important and those parameters discussed here.

2.2 RADIATED AND RECEIVED POWER

2.2.1 Radiated Power

For a homogeneous, isotropic, linear and lossless dielectric medium, the magnetic vector potential A is obtained from the Maxwell equations [115], [173]. The free space wave number is represented as:

$$A = \frac{\mu}{4\pi} \int_v \frac{J}{R} e^{-jkR} dv' \qquad (2.1)$$

here,

μ is magnetic permeability

J is electric current density

R is Relative distance between source point and field point= $|r - r'|$

k is wave number = $\omega\sqrt{\mu_0 \epsilon_0}$

$\qquad\qquad\qquad = \omega/c$

$\qquad\qquad\qquad = 2\pi/\lambda$; ($\lambda$ is wave length of the medium)

Assuming a Hertzian dipole source the current i(t) is written as:

$$i(t) = \text{Re}(Ie^{jkR}) \qquad (2.2)$$

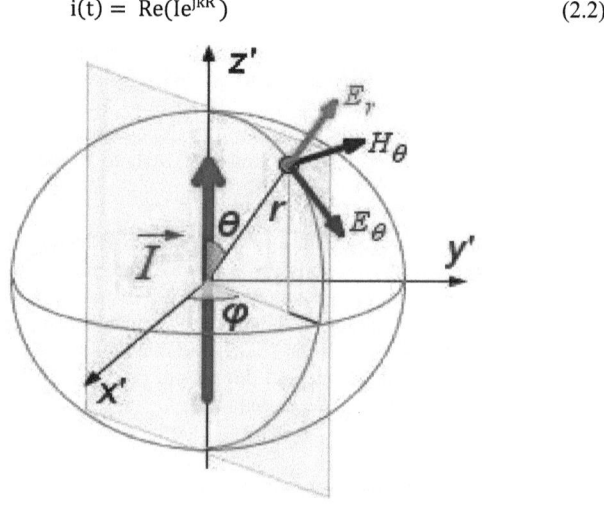

Figure 2.1 The Hertzian Dipole

Considering figure 2.1 and replacing J dv' in equation (2.1) with $\hat{z}I\,dz'$, A is further written as:

$$A = \frac{\mu}{4\pi}\int_v \frac{1}{R}e^{-jkR}\,dz'\hat{z} \tag{2.3}$$

so,
$$A = \frac{\mu Il}{4\pi r}e^{-jkr}\hat{z} \tag{2.4}$$

This expression suggests that the wave is propagating radially in the directions of $\hat{r}$ having phase constant k. The amplitude of the wave is inversely proportional to the distance [199]. The magnetic flux density B and magnetic vector potential A are related as:

$$B = \mu H = \Delta \times A \tag{2.5}$$

For spherical coordinate System the $\hat{z}$ is defined as:

$$\hat{z} = (\cos\theta\hat{r} - \sin\theta\,\hat{\theta}) \tag{2.6}$$

From equation (2.4) on substituting $\hat{z}$, now A is further expressed as:

$$A = \frac{\mu Il}{4\pi r}e^{-jkr}(\cos\theta\hat{r} - \sin\theta\hat{\theta}) \tag{2.7}$$

with the help of Maxwell equation the expression for the magnetic field intensity

$$H = \frac{1}{\mu}[\Delta \times A] = \frac{jkIl}{4\pi r}\sin\theta\left(1 + \frac{1}{jkr}\right)e^{-jkr}\hat{\phi} \tag{2.8}$$

For the electric field far away from the dipole, Maxwell's equation for J=0 yields,

$$E = \frac{1}{j\omega\varepsilon}[\nabla \times H] \tag{2.9}$$

Now using equation (2.8), E is further expressed as:

$$E = \frac{\eta Il}{2\pi r^2}\cos\theta\left(1 + \frac{1}{jkr}\right)e^{-jkr}\hat{r} + \frac{jk\eta Il}{4\pi r}\sin\theta\left(1 + \frac{1}{jkr} - \frac{1}{k^2r^2}\right)e^{-jkr}\hat{\theta} \tag{2.10}$$

where the intrinsic impedance of free space is introduced as:

$$\eta = \sqrt{\frac{\mu_0}{\varepsilon_0}} \tag{2.11}$$

Its value is 120π for free space.

For the near field region (kr<<1) the instantaneous Poynting vector which corresponds to the vector power density (W/m^2) is given as:

$$w(t) = e(t) \times h(t) = \text{Re}\{Ee^{j\omega t}\} \times \text{Re}\{He^{j\omega t}\} \tag{2.12}$$

By writing the real magnetic and electric field as:

$$h(t) = \frac{1}{2}[He^{j\omega t} + H^*e^{-j\omega t}] \tag{2.13}$$

$$e(t) = \frac{1}{2}[Ee^{j\omega t} + E^*e^{-j\omega t}] \tag{2.14}$$

the instantaneous Poynting vector can be rewritten as:

$$w(t) = e(t) \times h(t) = \frac{1}{2}\text{Re}\{[E \times H]e^{j2\omega t} + [E \times H^*]\} \tag{2.15}$$

For the near field i.e $kr \ll 1$ yields,

$$<W_r(t)> = \frac{1}{2}\text{Re}\left\{-j\frac{|I|^2 l^2}{16\pi^2 r^3 \omega \epsilon_0}\sin^2\theta \hat{r}\right\} = 0 \tag{2.16}$$

In equation (2.16), the $-j$ indicates that the near zone has capacitive behaviour which means that the dominant field is purely reactive and hence has zero average power [115], [173].

In the other case, assuming that the observation point is far away from the dipole ($kr \gg 1$), the terms $1/r^2$ and $1/r^3$ get extremely small. The far field components then become,

$$H \approx \frac{jkIl\sin\theta}{4\pi r}e^{-jkr}\hat{\phi} \tag{2.17}$$

and

$$E = \frac{jk\eta Il\sin\theta}{4\pi r}e^{-jkr}\hat{\theta} \tag{2.18}$$

From these equations, it is seen that the far field is a spherical wave with H_ϕ and E_θ field which are perpendicular, transverse and propagating in the r direction. In this case the medium's intrinsic impedance can be written as,

$$\eta = \frac{E_\theta}{H_\phi}, \tag{2.19}$$

For the far field, time averaged vector power density is

$$= \frac{1}{2}\text{Re}\{E \times H^*\} = \frac{1}{2\eta}|E_\theta|^2\hat{r} \tag{2.20}$$

$$W(t) = \frac{|I|^2 l^2}{32\pi^2 r^2}k^2\eta\sin^2\theta\hat{r} \quad (\text{W/m}^2). \tag{2.21}$$

The power density in the far field is purely real and directed radially outwards. Thus this is also called the radiated power per unit area. The total radiated power then becomes,

$$P_{rad} = \oint_s \{w(t)\}\, ds \qquad (2.22)$$

Substituting equation (2.21) into (2.22) and integrating over a large sphere P_{rad} is given as:

$$P_{rad} = \frac{|I|^2 l^2}{32\pi^2} k^2 \eta \int_0^\pi \sin^3 \theta \int_0^{2\pi} d\phi = \frac{\eta}{12\pi} k^2 |I|^2 l^2 \qquad (2.23)$$

In free space ($\eta = 120\pi$)

$$P_{rad} = 40\pi^2 \left(\frac{l}{\lambda}\right)^2 |I|^2 \qquad (2.24)$$

2.2.2 Radiation Resistance and Received Power [84]

The relation between current and total radiated power from dipole for a far field is purely real, the far field can linked to a resistance R_{rad} called radiation resistance,

$$P_{rad} = \tfrac{1}{2}|I|^2 R_{rad}. \qquad (2.25)$$

Thus

$$R_{rad} = \frac{2}{|I|^2} P_{rad}, \qquad (2.26)$$

And in free space,

$$R_{rad} = 80\pi^2 \left(\frac{l}{\lambda}\right)^2. \qquad (2.27)$$

The scenario in Figure 2.2 assumes that the antennas are lossless Hertzian dipoles ($R_{loss} = 0$) polarized in z direction. The electric field at distance r is,

$$E_\theta = \frac{jk\eta Il \sin\theta}{4\pi r} e^{-jkr} \qquad (2.28)$$

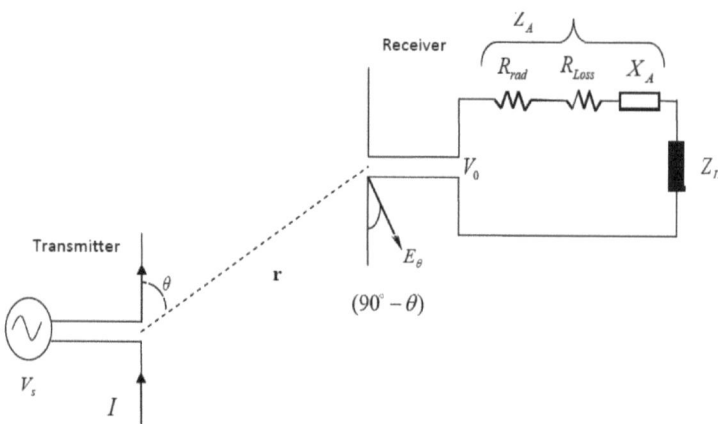

Figure 2.2 Voltage Induced at the Receiver Antenna

The electric field impinges on the receiving antenna at an angle and induces voltage proportional to E_0 and l. So the induced voltage is,

$$V_0 = E_0 l \sin\theta. \qquad (2.29)$$

here,

l is length of the receiver antenna and E_0 is vertical component of the electric field.

From equation (2.18), E_0 can be written as:

$$E_0 = \frac{jk\eta I l}{4\pi r}\sin\theta \qquad (2.30)$$

To find out the average power delivered to the load in Figure 2.2 assume the load is matched to the antenna. That means the antenna impedance $Z_A = Z_L^*$ [115]. The total impedance is $2R_{rad}$ yields maximum power to the load according to Jakobi's law. The power delivered to the load becomes,

$$P_r = \frac{1}{2}[V_0/2R_{rad}]^2 R_{rad} = \frac{1}{8R_{rad}}E_0^2 l^2 \sin^2\theta. \qquad (2.31)$$

2.2.3 Friis Transmission Equation

In antenna theory, the receiving antenna's power capturing capability is represented as:

$$A_{er} = \frac{P_r}{\langle W(t)\rangle} \qquad (2.32)$$

It is defined as the ratio of average power received by the antenna's load, to the time average power density at the antenna, and is called effective area. Earlier the average power density for the far-field was introduced as:

$$|\langle W(t)\rangle| = \frac{|E_0|^2}{2\eta}. \qquad (2.33)$$

From the equations (2.31), (2.32) and (2.33) the effective area of the receiving antenna can be written as:

$$A_{er} = \frac{\eta}{4R_{rad}} l^2 \sin^2\theta \qquad (2.34)$$

by using equation (2.26) the above equation becomes,

$$A_{er} = \frac{\lambda^2}{4\pi} 1.5\sin^2\theta = \frac{\lambda^2}{4\pi} G_r \qquad (2.35)$$

where,

$$G_r = 1.5\sin^2\theta \qquad (2.36)$$

is the directive gain of the Hertzian dipole. Equation (2.35) shows that the receiving antenna's effective area is independent of its length and inversely proportional to the square of the carrier frequency [115], [173]. At this point one can realize that the term frequency dependent propagation loss is not the affect of wave propagation but the receiving antenna itself. The average power density in terms of radiated power, transmitter gain G_t, and distance r is written as:

$$\langle W(t)\rangle = \frac{P_{rad}G_t}{4\pi r^2}. \qquad (2.37)$$

considering,

$$P_r = <W(t)> A_{er} \qquad (2.38)$$

the equations (2.35), (2.37) and (2.38) yield the following formula for P_r

$$P_r = P_{rad} G_t G_r \left(\frac{\lambda}{4\pi r}\right)^2 \qquad (2.39)$$

It is Friis transmission formula and it gives a relation between the power radiated by the transmitting antenna and the power received by the receiving antenna. The Path loss for the free space in dB then can be written as follows:

$$PL(dB) = 10\log\frac{P_{rad}}{P_r} = -10\log\left[\frac{G_t G_r \lambda^2}{(4\pi)^2 (r)^2}\right] \qquad (2.40)$$

The propagation path loss is discussed in the following section in detail.

2.3. PROPAGATION PATH LOSS

In wireless communication systems, the information is transmitted from transmitter to receiver by means of electromagnetic waves [196]. Signal strength of the electromagnetic wave is reduced due to the interface between the electromagnetic waves the channel (environment or free space or atmosphere). The pictorial representation of the path loss occurrence is given in figure 2.3. Different models are there to calculate the path loss obtained in a communication channel [4]. The power of signal is decreased due to path distance, reflection, diffraction, scattering, and free-space loss by the physical objects of environment [31]. Variations of transmitter and receiver antenna heights also create losses. In general path loss is expressed as:

$$PL = \frac{Power Transmitted}{Power \operatorname{Re}ceived}$$

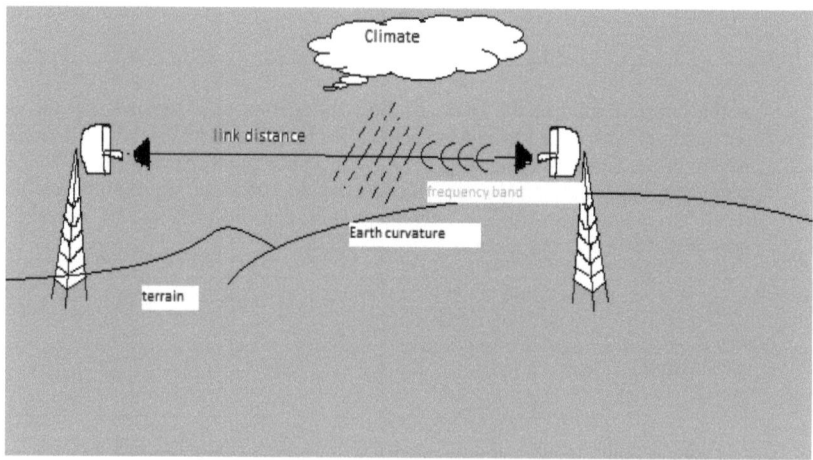

Figure 2.3 Illustration of Wireless Communication Showing Path Loss

2.3.1 Causes of Path Loss

Signal path loss is caused by various factors. In a global environment, the actual RF path loss is affected by many factors [118]. When scheduling any radio or wireless system, it is essential to have a broad perceptive about the elements that leads to the path loss [116]. The following are some of the major elements that cause signal path loss for any wireless radio communication system [171].

- ➤ Free space loss: It is a general loss occurs when a signal travels from transmitter to receiver through space without any other effects attenuating the signal. According to the conversion of energy, signal energy reduces when it travels a larger distance in the space.

- ➤ Absorption losses: This loss occurs when a radio signal passes into a medium like large buildings and foliage which are not totally transparent to radio signals then these types of losses are occurred. The environment over which signals pass through will have a major effect on the signal. The hills obstruct the path and considerably attenuate the signal and again making reception impossible. It is found that signals travel better over more conductive terrain, e.g. sea paths. Higher levels of attenuation are obtained from dry sandy terrain. Buildings and other obstructions including vegetation have a significant effect. The buildings reflect radio signals and also absorb them. Trees and foliage can attenuate radio signals, particularly when they are wet.

- ➤ Diffraction: This type of losses occurs when an obstruction unexpectedly comes in the path. At sharp edges, the radio signals are more diffracted [190].

- ➤ Multipath: In general the signals which are reflected in the channel will reach the receiver via a number of different paths. Depending upon the relative phases of the signals, they may be added and subtracted from each other. This entire process leads to a loss which is called multipath loss.

- ➤ Atmosphere: The radio signal paths are also affected by the atmosphere. The ionosphere has a major effect at lower frequencies (below 30 - 50MHz). At frequencies above 50 MHz the troposphere has a major effect on the radio signal path. For UHF broadcast this can enlarge coverage to approximately a third beyond the horizon.

To design a path loss propagation model it is necessary to understand mobile radio propagation environment and it is discussed here in the fallowing section.

2.4 MOBILE RADIO PROPAGATION ENVIRONMENT

Electromagnetic wave propagates through a medium by reflection, refraction, diffraction and scattering. It depends on the wavelength compare to object sizes, inject angle of wave and atmospheric temperature [205].

2.4.1 Reflection

When electromagnetic wave propagates, it experiences a reflection due to object of the environment is large enough compared to its wavelength. Reflections are produced from many sources like the ground surfaces, the walls and from equipments as depicted in figure 2.4. The co-efficient of reflection and refraction mainly depend on angel of incidence, the operating frequency and the wave polarization [37].

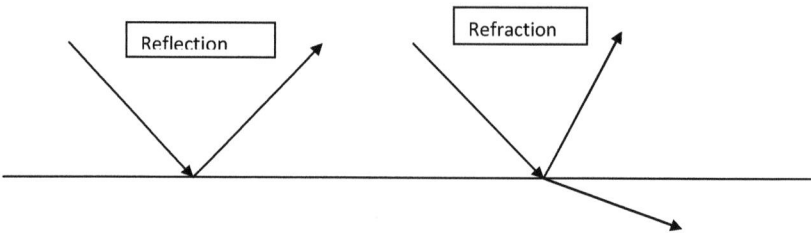

Figure 2.4 Phenomenon of Reflection and Refraction

2.4.2 Refraction

The density of atmosphere is affected by the variation of air temperature [37]. If a wave is incident upon this kind of atmospheric medium, the wave alters its direction from the original wave's path and leads to refraction. The refraction phenomenon is represented in figure 2.4.

2.4.3 Diffraction

Diffraction is created when the electromagnetic wave propagate from transmitter to receiver obstructed with a sharp edge surface as presented in figure 2.5 [8]. When NLOS exist in the radio path, the diffraction occurs when wave propagates behind the obstacle. Not only the geometry of the object, but also the angle of incident, amplitude and phase of the signal are responsible for making diffraction [52].

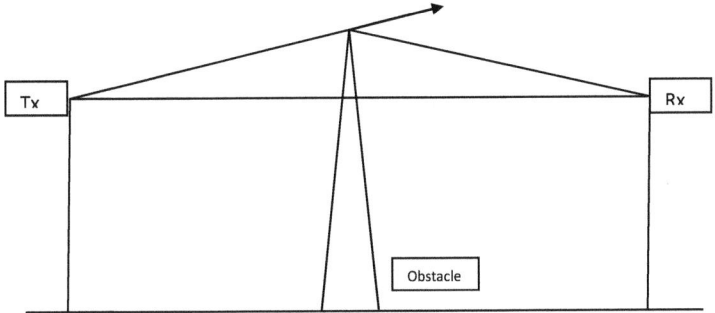

Figure 2.5 Diffraction in Sharp Edge

2.4.4 Scattering

If the object of the environment is enough small compared to the wavelength of the signal or the number of obstacles per unit is enough large compared to the signal, than scattering occurs as depicted in figure 2.6. In the practical field, it occurs due to small objects like foliage, lampposts and street signs especially in the city area.

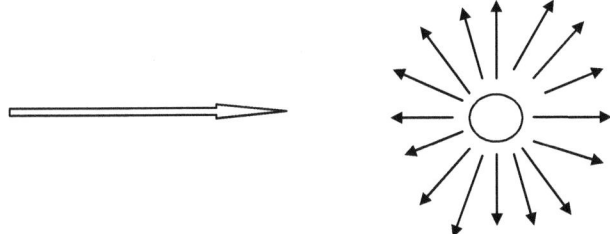

Figure 2.6 Wave is Scattered by a Small Obstacle

Existing Field propagation path loss models study is also very important in this research. So it is discussed in the following section.

2.5 FIELD PROPAGATION PATH LOSS MODELS

Path loss prediction model or Field propagation models are derived by combining analytical and empirical methods. These models are used for estimating the electromagnetic field strength at network planning level of a mobile radio system. It describes the signal reduction from transmitter to receiver antenna as a function of distance, carrier frequency, antenna heights and other significant parameters like terrain profile. Several models are used for estimating initial deployment and those can be generally categorized into outdoor and indoor propagation model [74]. The outdoor propagation models are used to understand the link performance of macro cellular systems whereas indoor model are used for micro cellular environment [21].

2.5.1 Indoor Field Propagation Models

With the advent of Personal Communication System (PCS), there is an immense deal of interest in characterizing electromagnetic wave propagation inside buildings. There are two aspects, which differ the indoor radio channel and the traditional mobile radio channel – the distance covered is smaller range, and the variability of the environment is much superior for small range of T-R separation distances [97]. It has been analysed that propagation within buildings is strongly influenced by specific features like the layout of the building, the construction materials, and the building type. Indoor radio propagation is dominated by the same mechanism as outdoor: reflection, diffraction, and scattering. For example, signal levels fluctuate greatly depending on whether interior doors are opened or closed inside the building [112]. Where antennas are mounted also impacts large-scale propagation. Antenna mounted at desk level in a partitioned office receives enormously different signals than those mounted on the ceiling. Also, the smaller propagation distances make it more difficult to assure far-field radiation for all receiver location and type of antennas [23]. In general, indoor channels may be classified either as line of sight (LOS) or obstructed, with varying degrees of clutter. The key models [203] which have recently emerged for indoor are:-

- Log distance Path Loss Model
- Ericsson Multiple Breakpoint Model
- Attenuation factor model

2.5.2 Outdoor Field Propagation Models[21]

2.5.2.1 Empirical Models

These models are used where mathematical model analysis is not feasible to explain a random situation. In that case, these models use some data to predict the approximate behaviour of the situation. By definition, an empirical model is based on data used to predict, not to explain a system and are based on observations and measurements alone. It can be divided into two subcategories as time dispersive and non-time dispersive. The time dispersive model gives us with information about time dispersive characteristics of the channel like delay spread of the channel during multipath. The Stanford University Interim (SUI) model is the ideal example of this type. COST 231 Hata model, Hata and ITU-R model are example of non-time dispersive empirical model.

2.5.2.2 Deterministic Models

This makes use of the laws governing electromagnetic wave propagation in order to find out the received signal power in a particular location. Nowadays, the visualization capabilities of computer increases quickly. The modern systems of predicting radio signal coverage are Site Specific (SISP) and uses Graphical Information System (GIS) database. SISP model can be connected with indoor or outdoor propagation environment as a deterministic type. Wireless system designers are capable to design actual presentation of buildings and terrain features by using the building databases. For indoor propagation models, architectural drawing gives a SISP representation. Wireless systems have been developing the use of computerized design tools that certify more deterministic comparing statistical.

2.5.2.3 Stochastic Models

This is used to represent the environment as a series of random variables. Least information is required to sketch this model but its accuracy is questionable. Prediction of electromagnetic wave propagation at 3.5GHz frequency band is derived by combining empirical and stochastic models.

2.6 EMPIRICAL MODEL

A radio wave propagation model is a sequence of mathematical calculations derived to predict signals characteristics and the associated losses in a given environment. The prophecy is based on some parameters such as frequency, distance and the clutter type. The propagation models are empirical in nature since they are derived from the real data sets. The collected data is analysed and formulas are developed to fit the data curves. The model is based on composed data determined by the formulas limited to specific ranges. When a variety of complex conditions exist then these developed models help to predict the signal characteristics. Modelling allows the use of approximation methods to account the propagation losses by varying the signal parameters. Some outdoor path loss propagation models (Empirical) are discussed below [227]:

2.6.1 Free Space Path Loss Model

Friis developed the basic Free Space Propagation model and can be applied for any frequency [119]. This model is used to predict the received signal strength from the transmitting antenna with an unobstructed Line Of Sight (LOS) between the antennas [199]. The model does not include the affects of absorbing obstacles, reflecting surfaces and influence of earth surface. This model is the simplest formula of the ratio between transmitted power and received power. Friis equation is mentioned in equation 2.44.

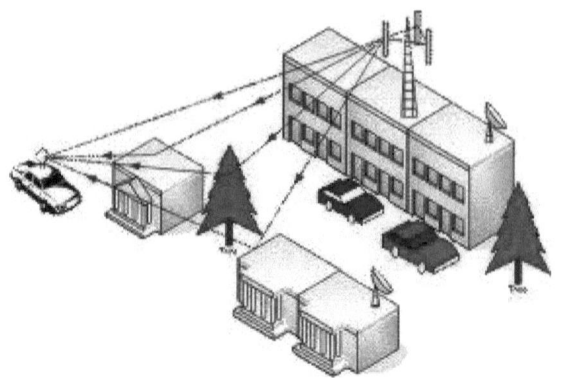

Figure 2.7 Example of Free Space Communication

$$P_r(d) = \frac{P_t G_t G_r \lambda^2}{(4\pi)^2 d^2 \, L} \qquad (2.44)$$

where,
P_t is transmitter power,
G_t is transmitter antenna gain,
G_r is receiver antenna gains,
L is loss factor of the system not related to propagation (L ≥ 1),
λ is signal wavelength in meters
f is frequency in MHz
d is distance in km

The gain of an antenna is related to its effective aperture, A_e by

$$G = \frac{4\pi A_e}{\lambda^2} \qquad (2.45)$$

The effective aperture A_e is related to the physical size of the antenna. The path loss for the free space model in dB is

$$PL(dB) = -10 \log \frac{P_r}{P_t} = -10 \log \frac{G_t G_r \lambda^2}{(4\pi)^2 d^2} \qquad (2.46)$$

When the antenna are assumed to have unity gain, and the path loss is expressed as:

$$PL(dB) = -10\log\frac{P_r}{P_t} = -10\log\frac{\lambda^2}{(4\pi)^2 d^2} \qquad (2.47)$$

Or
$$PL\ (dB) = 32.44 + 20\log f + 20\log d \qquad (2.48)$$

The above equation shows that losses increase with distance as well as frequency.

2.6.2 Lee Path Loss Model

Lee's path loss model is also based on empirical data (flat terrain). This model is good for flat terrains not affective for other terrains. However, Lee's model has been recognized to be more of a "North American model" than that of Hata [99], [102]. The propagation loss calculated as:

$$PL\ (dB) = 124 + 30.5 \log_{10}\left(\frac{D}{D_0}\right) + 10k \log_{10}\left(\frac{f}{f_c}\right) - \alpha_0 \qquad (2.49)$$

here
- $k = 2$ for $f_c < 450$ MHz   in suburban/open area
- $k = 3$ for $f_c > 450$ MHz   in urban area
- $D_0 = 1.6$ km.

f is transmitted frequency in MHz,
D is Transmitter- Receiver distance in Km and
α_0 is correction factor to account BS and MS antenna heights.

$$\alpha_0 = \alpha_1 \alpha_2 \alpha_3 \alpha_4 \alpha_5 \qquad (2.50)$$

where

$$\alpha_1 = \left(\frac{new\ BS\ antenna\ height(m)}{30.48(m)}\right)^2$$

$$\alpha_2 = \left(\frac{new\ MS\ antenna\ height(m)}{3(m)}\right)^g$$

$$\alpha_3 = \left(\frac{new\ transmitter\ Power}{10W}\right)^2$$

$$\alpha_4 = \left(\frac{new\ BS\ antenna\ gain\ with\ respect\ to\ \lambda_c/2\ dipole}{4}\right)$$

$\alpha_5 =$ Different antenna gain correction factor at MS   (2.51)

The value of n and Ɛ are also based on empirical data and recommended to take the following values:

$$n = \begin{pmatrix} 2.0 & \text{for} & f_c < 450 MHz & \text{Rural and Suburban} \\ 3.0 & \text{for} & f_c > 450 MHz & \text{Urban} \end{pmatrix} \quad (2.52)$$

$$\xi = \begin{pmatrix} 2.0 & \text{for} & \text{MS antenna height} > 10m \\ 3.0 & \text{for} & \text{MS antenna height} < 3m \end{pmatrix} \quad (2.53)$$

2.6.3 COST 231 Walfish-Ikegami (W-I) Model

This model is a combination of J. Walfish and F. Ikegami model [53]. This model is the most appropriate for flat suburban and urban areas that have uniform building height. Among the all other models, COST 231 W-I model gives a more precise path loss. This is as a result of the additional parameters introduced which characterised the different environments. It distinguishes terrain with different proposed parameters [197].

In case of LOS, path loss is defined as:
$$(PL)_{LOS} = 42.6 + 26\log_{10}(d) + 20\log_{10}(f) \quad (2.54)$$

And for NLOS condition
$$(PL)_{NLOS} = \{LFSL + Lrts + Lmsd\} \quad (2.55)$$

This is the extended version of COST-231 and consist rooftop losses and building losses.

2.6.4 Egli Path Loss Model

Egli model is very popular because of easy implementation and concurrence with the empirical data for analysis [7]. Egli path loss model equation for path loss over irregular terrain is:

$$PL (dB) = G_B G_M \left(\frac{h_B h_M}{d^2}\right)^2 \left(\frac{40}{f}\right)^2 \quad (2.56)$$

here,
  G_B is gain of base station antenna
  G_M is gain of mobile station antenna
  h_B is height of base station antenna in meters
  h_M is height of mobile station antenna in meters
  d is distance from base station antenna in km and
  f is frequency of transmission in MHz

Egli model provides the entire path loss, where as the Okumura model discussed below provides the path loss in addition to free space loss. This model is empirical in nature and valid to scenarios where the transmission has to pass through an irregular terrain. Egli model is not applicable to such situations where some vegetative obstruction takes place in the middle of the link.

2.6.5 Okumura Model

Okumura model is the most extensively used model for signal prediction in the urban areas. This model is appropriate for frequency band from 150 MHz to 1920 MHz (although it is typically extrapolated up to 3000 MHz) and distances from 1 km to 100 km. It is used for base station antenna heights ranging from 30 m to 1000 m [7]. Okumura developed a set of median attenuation curves relative to free space (A_{mu}) in an urban area over quasi smooth terrain with a base station effective antenna height (h_{te}) of 200 m and a mobile antenna height (h_{re}) of 3 m. These curves were developed from broad measurement using vertical omni directional antennas at both the base station and mobile station, and are plotted as a function of frequency ranging from 100 MHz to 1920 MHz and also as a function distance from base station in the range 1 km to 100 km. To determine path loss using Okumura model, the free space path loss is first determined, and then the value of A_{mu}(from curve) is included to it having correction factors to account for the terrain type [94]. The Okumura Model is expressed as:

$$PL(dB) = L_F + Amu(f,d) - G(h_{te}) - G(h_{re}) - G_{area} \qquad (2.57)$$

here PL(dB) is the propagation path loss, L_F is free space path loss, A_{mu} is the median attenuation relative to free space attenuation (shown in the figure 2.9), $G(h_{te})$ is base station antenna height gain factor, $G(h_{re})$ is mobile antenna height gain factor, and G_{area} is gain corresponding to specific environment (shown in the figure 2.8).

$G(h_{te})$ and $G(h_{re})$ is expressed as:

$$G(h_{te}) = 20\log_{10}\left(\frac{h_{te}}{200}\right) \quad 1000m > h_{te} > 30m \qquad (2.58)$$

$$G(h_{re}) = 10\log_{10}\left(\frac{h_{re}}{3}\right) \quad h_{re} \leq 3m \qquad (2.59)$$

$$G(h_{re}) = 20\log_{10}\left(\frac{h_{re}}{3}\right) \quad 10m > h_{re} > 3m \qquad (2.60)$$

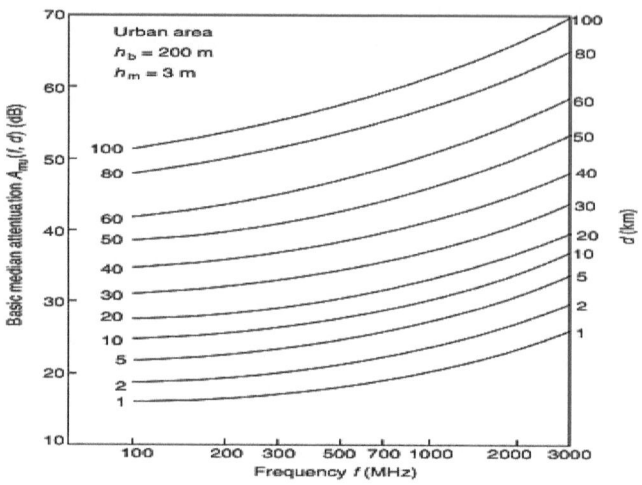

Figure 2.8 Median Attenuation Relative to Free Space $A_{mu}(f,d)$ Over a Quasi-smooth Terrain

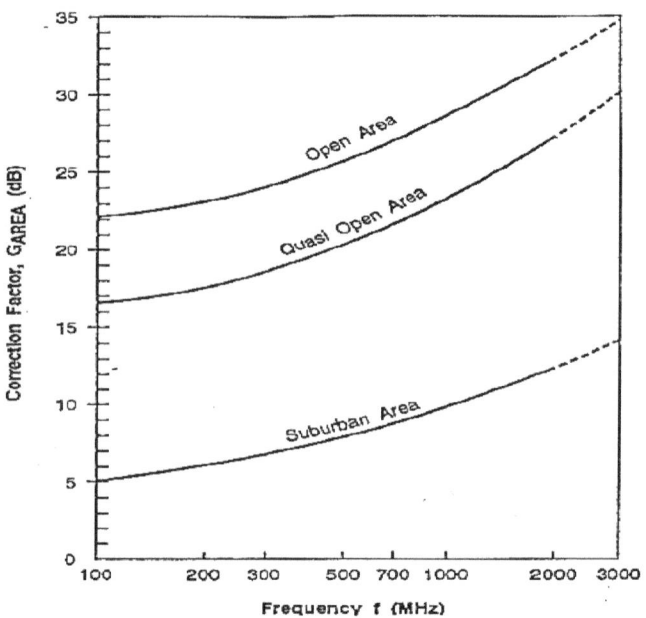

Figure 2.9 Correction Factor G_{area} for Different Types of Terrain

2.6.6 Hata Model

Hata created a mathematical formulation from Okumura prediction curves in order to acquire simple computational applications [7]. Thus, this model is then called Okumura-Hata model or simply Hata model. Okumura prediction model is based on propagation measurement conducted in Kanto (near Tokyo), Japan by Okumura. This model presents signal strength prediction curves over distance in a quasi-smooth urban area (terrain fold is less than 20 m). In order to predict other types of area classifications, correction factors for suburban and open areas are given. Okumura also gives correction factors for different terrain irregularities, such as hilly terrain, mixed land-sea path, and diffraction by ridges [122]. With the help of Okumura prediction model, the parameters such as base station antenna height, terrain undulation height, terrain slope, etc. should be determined according to the Okumura report. Okumura-Hata model for path loss prediction in urban, suburban, and open area are given as:

$$PL_{po(OH)}(dB) = PL_{pu(OH)} - 4.78\,(\log f_c)^2 + 18.33 \log f_c - 40.94 \qquad (2.61)$$

$$PL_{ps(OH)}(dB) = PL_{pu(OH)} - 2\,\{\log\,(f_c/28)\}^2 - 5.4 \qquad (2.62)$$

$$PL_{pu(OH)}(dB) = 69.55 + 26.16 \log f_c - 13.82 \log h_{te} - a(h_m) \\ + (44.9 - 6.55 \log h_{te})\log d \qquad (2.63)$$

where,
$PL_{po(OH)}$ is path loss prediction for open area in dB
$PL_{ps(OH)}$ is path loss prediction for suburban area in dB
$PL_{pu(OH)}$ is path loss prediction for urban area in dB
f_c is carrier frequency in MHz
d is distance base station and mobile station in km
h_{te} and h_{re} are base station and mobile antenna heights, respectively in meters
$a(h_m)$ is correction factor of mobile antenna height in dB

In equation of urban path loss, $a(h_m)$ is defined for two different areas, medium or small city and large city areas. In a medium or small city area,

$$a(h_m)\,[dB] = (1.1 \log f_c - 0.7)h_{re} - (1.56 \log f_c - 0.8) \qquad (2.64)$$

While in a large city area,

$$a(h_m)\,[dB] = 8.29\,(\log 1.54\,h_{re})^2 - 1.1 \text{ for } f_c \geq 300 \text{ MHz} \qquad (2.65)$$

$$a(h_m)\,[dB] = 3.2\,(\log 11.75\,h_{re})^2 - 4.97 \text{ for } f_c \leq 300 \text{ MHz} \qquad (2.66)$$

2.6.7 COST 231 Model

This model is widely used for predicting path loss in mobile wireless system. It is an extension of Hata model. The COST 231 Hata model is designed for the frequency band in between 1500 MHz to 2000 MHz. It also consist corrections for urban, suburban and rural (flat) environments [48], [63]. The basic equation for path loss in dB is given below:

$$PL(dB) = 46.3 + 33.9 \log_{10}(f) - 13.82 \log_{10}(h_{te}) - a(h_m) + (44.9 - 6.55 \log_{10}(h_{te})) \log_{10}d + c_m \quad (2.67)$$

where, f is the frequency in MHz, d is the distance in between base station and mobile antennas in km, and h_{te} is the base station antenna height measured from ground level in meters. The c_m is defined as 0 dB for suburban or open environments and 3 dB for urban environments. The parameter $a(h_m)$ is defined for large area as:

$$a(h_m) [dB] = 8.29 (\log 1.54 h_{re})^2 - 1.1 \quad \text{for } f \geq 300 \text{ MHz} \quad (2.68)$$

$$a(h_m) [dB] = 3.2 (\log 11.75 h_{re})^2 - 4.97 \quad \text{for } f \leq 300 \text{ MHz} \quad (2.69)$$

And for medium or small city area,

$$a(h_m) [dB] = (1.1 \log_{10}f - 0.7)h_{re} - (1.56 \log_{10}f - 0.8) \quad (2.70)$$

here, h_{re} is the mobile antenna height above ground level in meters.

2.6.8 ECC-33 Path Loss Model

The Okumura experimental data was collected in the suburban areas of Tokyo. This model well suited to urban areas subdivided into 'large city' and 'medium city' groups. They also provide correction factors for 'suburban' and 'open' areas. Since the characteristics of a extremely developed area such as Tokyo are quite different to those which are found in typical European suburban areas, use of the 'medium city' model is recommended for European cities. Although the Okumura-Hata model is mostly used for UHF bands, its accuracy is uncertain for higher frequencies. The COST 231Hata model extended its use up to 2 GHz but it was proposed for mobile systems having Omni directional CPE antennas sited less than 3 m above ground level. A different approach was taken in which the original measurements by Okumura and customized its assumptions so that it more closely represents a fixed wireless access system. The path loss model presented is referred here as the ECC-33 model. The path loss expression is given as,

$$PL(dB) = A_{fs} + A_{bm} - G_b - G_r \quad (2.71)$$

where, A_{fs}, A_{bm}, G_b and G_r are the free space attenuation, the basic median path loss, the BS height gain factor and the terminal (CPE) height gain factor. They are individually defined as,

$$A_{fs} = 92.4 + 20 \log_{10}(d) + 20 \log_{10}(f) \quad (2.72)$$

$$A_{bm}= 20.41 + 9.83 \log_{10}(d) + 7.894 \log_{10}(f) + 9.56[\log_{10}(f)]^2 \qquad (2.73)$$

$$G_b= \log_{10}(h_b/200)\{13.958 + 5.8[\log_{10}(d)]^2\} \qquad (2.74)$$

and for medium city environments,

$$G_r = [42.57 + 13.7 \log_{10}(f)][\log_{10}(h_r) - 0.585] \qquad (2.75)$$

where, f is the frequency in GHz, d is the distance between base station and CPE in km, h_b is the BS antenna height in meters and h_r is the CPE antenna height in meters. The medium city model is more suitable for European cities whereas the large city environment should be used for cities having large buildings. It is interesting to note that the predictions produced by the ECC-33 model do not lie on straight lines when plotted against distance having a log scale.

2.6.9 Bullington Model

This model approximates techniques to calculate the diffraction loss over multiple knife edges. A new 'effective' obstacle is defined by this model at the point where the line-of-sight from two antennas across. There are many practical applications in urban and rural areas to the model [96].

2.6.10 Epstein-Peterson Model

This model is similar to the Bullington model with the exception that it suggests to draw line-of-sight between appropriate obstacles, and to adjoin the diffraction losses at each obstacle [137]. However this method does not take urban losses into account, and 10 dB or more must be added to the calculated loss in urban areas.

2.6.11 Stanford University Interim (SUI) Model

SUI path loss prediction model has been developed under the Institute of Electrical and Electronic Engineers (IEEE) 802.16 Broadband Wireless Access Working Group. This prediction model is the extension of Hata model with frequency larger than 1900 MHz. The correction parameters are certified to enlarge this model up to 3.5 GHz band. In the USA, this model is used for the Multipoint Microwave Distribution System (MMDS) for the frequency band from 2.5 GHz to 2.7 GHz. The base station antenna height of SUI model can be varied from 10 m to 80 m. Receiver antenna height is ranging from 2 m to 10 m. The cell radius is from 0.1 km to 8 km. The SUI model elaborates three types of terrain namely terrain A, terrain B and terrain C. There is no assertion about any particular environment. Terrain A can be used for hilly areas with moderate or very dense vegetation. This terrain gives the highest path loss.

Terrain B is characterized for the hilly terrains with rare vegetation, or flat terrains with moderate tree densities. This terrain can be considered for suburban environment. Terrain C is suitable for flat terrains or rural with light vegetation. The basic path loss expression of The SUI model is given below along with correction factors.

$$PL(dB) = A + 10\gamma \log_{10}\left(\frac{d}{d_0}\right) + X_f + X_h + s \qquad \text{for } d > d_0 \qquad (2.76)$$

where,
  d is distance between BS and receiving antenna in meter
  d_0 is 100 m
  λ is Wavelength in meter
  X_f is Correction for frequency above 2 GHz
  X_h is Correction for receiving antenna heights
  s is Correction for shadowing in dB
  γ is Path loss exponent

The log normally distributed factor s, for shadow fading because of trees and other clutter on a propagations path and its value is between 8.2 dB and 10.6 dB. The parameter A is defined as:

$$A = 20 \log_{10}\left(\frac{4\pi d_0}{\lambda}\right)$$

and the path loss exponent γ is given by

$$\gamma = a - b h_b + \left(\frac{c}{h_b}\right)$$

where, the parameter h_b is the height of base station antenna in meters. This is between 10 m and 80 m. The constants a, b, and c entirely depend upon the types of environment, that are given in table 2.1. The value of parameter $\gamma = 2$ for free space propagation in an urban area, $3 < \gamma < 5$ for urban NLOS environment, and $\gamma > 5$ for indoor propagation.

Table 2.1 The Parameter Values of Different Terrain for SUI Model.

Model Parameter	Terrain		
	A	B	C
A	4.6	4	3.6
b (m^{-1})	0.0075	0.0065	0.005
c(m)	12.6	17.1	20

The frequency correction factor X_f and the correction for receiver antenna height X_h for the model are expressed as:

$$X_f = 6.0 \log_{10}\left(\frac{f}{2000}\right) \tag{2.77}$$

$$X_h = \begin{cases} -10.8 \log_{10}\left(\dfrac{h_r}{200}\right) & Terrain Type A \\ -20.0 \log_{10}\left(\dfrac{h_r}{200}\right) & Terrain Type C \end{cases} \tag{2.78}$$

Where, f is the operating frequency in megahertz, and h_r is the receiver antenna height in meter. The correction factors given above increase the popularity of this model in all three types of terrain in rural, urban and suburban environments.

2.6.12 Walfisch- Bertoni Model

The impact of diffraction from rooftops and buildings has not been considered by the COST extension to the Hata model. For these effects Walfisch and Bertoni developed a model. It is used to predict average signal strength at street level using diffraction [30]. The model considers the path loss to be the product of three factors which is given as:

$$L = P_0 Q^2 P_1, \tag{2.79}$$

Where, P_0 is the free space path loss for omni directional antennas, Q^2 represents the signal power reduction due to buildings that block the receiver at street level, and P_1 is signal loss from the rooftop to the street due to diffraction. The model has been adopted for the IMT-2000 standard.

2.6.13 Longley Rice Model

When there is point to point communication systems take place then this model is applicable in the frequency range from 40MHz to 100GHz, over different kinds of terrain. This method operates in two modes. First is point to point mode prediction i.e., when a detailed terrain path profile exist. Second, If the terrain path profile is not available, this method provides techniques to determine the path specific parameters. Many modifications and corrections have been done to this model since its original publication. One deficiency of this model is that it does not provide a way of determining corrections due to environmental factors in the immediate vicinity of the mobile receiver to account for the effects of buildings and foliage. Multipath is not considered in this model.

2.7 CONCLUSION

In this chapter, it has been surveyed different field propagation models to calculate path loss in different types of climatic conditions and terrains. The received signal strength from the base stations has been calculated to determine which model minimized the number of handoffs. The Hata and Okumura model has widespread applications in an open environment but is severely limited in built-up areas. The Walfisch-Ikegami model prevails well in city and suburban areas and therefore has the edge over the Hata-Okumura model. The Longley-Rice model is meticulous but accurate and has applications over very rough terrain, this is an advantage in certain circumstances. The Egli model though draggy can generally be used under most environmental conditions since it is a statistical model. As reference models the Bullington and Epstein-Peterson are readily available. It has been clearly seen that the selection of an appropriate propagation model determines how many base stations are required to provide coverage for a network in particular area in different climatic conditions.

CHAPTER 3

METHODLOGY FOR FIELD DATA COLLECTION, ANALYSIS & ITS SIMULATION IN MATLAB

In this chapter, total concentration is on data collection tools and its analysis because field data collection is one of the major requirements of current research work. A number of data collection tools are available in the market in which TEMS data collection tool has been chosen for the current research work because it provides several advantages as compared to other data collection tools.

3.1 INTRODUCTION

The excellence of the network is eventually evaluated by the satisfaction of the subscribers of the network. The measurement procedure is based on the use of a drive test system. A drive test simply means drive and test while roaming of a wireless network in a car. The drive test system provides the insight to the performance of a network particularly in terms of RF coverage. Drive tests give evaluation of the network performance the field. The testing procedure starts with selection of the area of the network where the drive tests require to be performed. The RF engineer must check the suitable kits that comprise mobile equipment (mobile phone), drive testing software (on a laptop), and a GPS (global positioning system) unit before going to the drive test. While the drive testing starts, two mobiles are used to make calls with a delay of few seconds. The third mobile is generally used for testing the coverage area. Using this third mobile the engineer makes one continuous call. If this call breaks it will go for another call. The main use of this testing process is to collect enough samples at a reasonable speed and in a reasonable time. It consists of investigating the RF coverage by looking at the key performance indicators, such as pilot power strength (Ec/Io), the forward transmit power (Tx), the downlink receive power (Rx) and the Frame Error Rate (FER) while roaming in the wireless network. In drive test evaluation, the BSs must be identified prior and then the route to be followed traced. This is very important in order to conduct a successful drive test in time [226]. The data collection tools used for drive test along with accessories is as follows:

- Drive test tool (TEMS)
- TEMS compatible mobile station with RF antenna and Serial Port data cable
- A laptop with minimum 256MB RAM
- Analysis software (Mapinfo, Actix etc.)

- Inverter
- Extension board
- GPS receiver having DC Power cable and Serial Port data cable.

3.2 DATA COLLECTION TOOLS

Drive Test is defined as a method that used to verify the real condition of RF signal for certain operator at certain place. The data collection tool is a product consisting of software and hardware components. Various test tools are available which are used for Field measurement, some of them are: - NEMO, TEMS, AGILENT. Now some features of each data collection tool are discussed here.

3.2.1 Nemo

Nemo is a drive test tool for measuring and monitoring the air interface of wireless networks with outdoor and indoor measurement options like voice-video-data quality and application testing option [165]. The window of user interface system is shown in figure 3.1. It can perform:

- Simultaneous testing of different network standards.
- Single and multi voice as well as data measurements.
- Automatic mode and manual mode for testing circuit and packet switched services with mobile phones.
- Fast pilot / frequency scanning with external scanner.

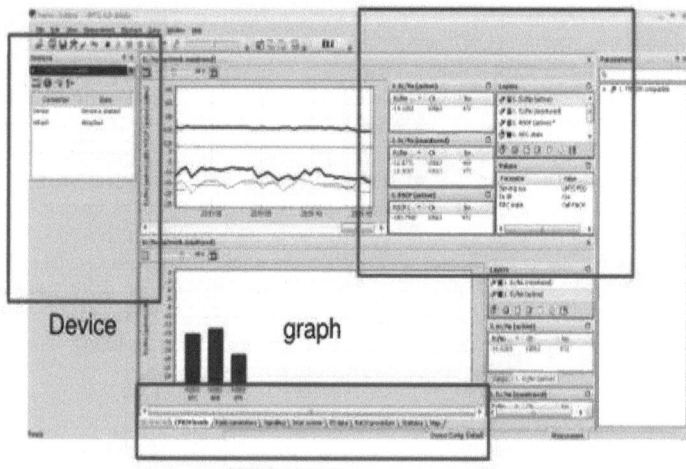

Figure 3.1 User Interface of Nemo Drive Test Tool

- MapX map support for MapInfo compatible digital maps.
- Nemo supports more than 40 test terminals on one common software platform.
- Nemo is compatible for all major wireless technologies: Global System for Mobile Communications (GSM), High-speed circuit-switched data (HSCSD), General packet radio service (GPRS), Enhanced Data Rates for GSM Evolution (EDGE), Wideband Code Division Multiple Access (WCDMA), High-Speed Downlink Packet Access (HSDPA), Advanced Mobile Phone System (AMPS), Time division multiple access (TDMA), code division multiple access (CDMA), CDMA2000 and Terrestrial Trunked Radio (TETRA).

3.2.2 Agilent

Agilent's GPRS and data test platform offers a true drive test solution. Agilent's drive test platform will allow the addition of four phones or a combination of phones and digital receivers to evaluate 2G, 2.5G and 3G networks simultaneously from the same laptop as shown in figure 3.2.

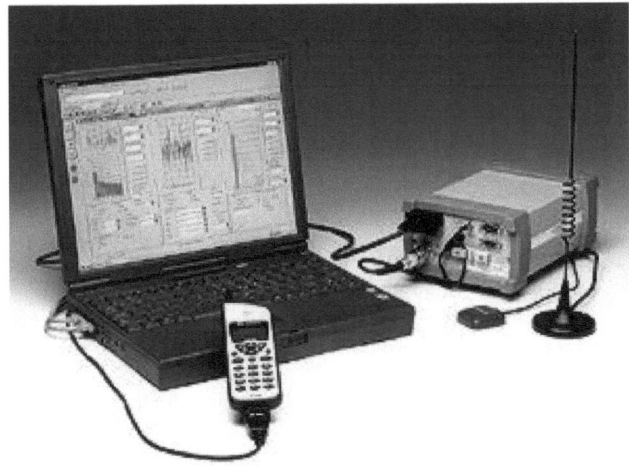

Figure 3.2 E7478A Drive Test System with E6455C IMT2000 Digital Receiver (Agilent Data Collection Tool)

The E7478A allows complete scalability on current formats (GSM, GSM-GPRS 900/1800/1900, TDMA, CDMA, W-CDMA/ Universal Mobile Telecommunications System (UMTS)). The features of E7478A are selection of a site, optimization, problem solving, end to end measurement of data, indoor analysis, and benchmarking.

3.2.3 Pioneer™

Pioneer is a software defined radio application for use with the 4301 radio platform. Pioneer enables the platform to perform network measurements in a mobile environment. It also provides an additional user interface to facilitate control and measurement recording for complete test system capability. The pioneer data collection tool is depicted in the following figure 3.3.

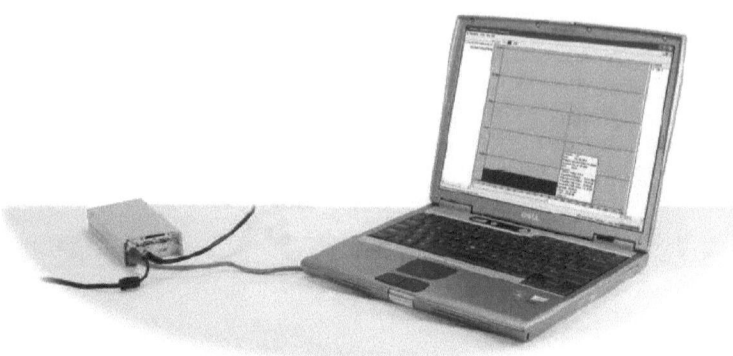

Figure 3.3 Pioneer Data Collection Tool

- Advanced measurement techniques and data monitoring for field network analysis and testing.
- The configuration of radio units and measurements is very simple.
- It is compatible for multiple 4301 units to allow flexible test solutions and configurations.
- It is user friendly and easy available application which support through DRT4000 application programming interface (API).
- It has a variety of data display capabilities and logging capabilities.
- It works as a Remote to laptop PC and local to unit log file support.
- It has 100 Mbps Ethernet interface to the host allows for high throughput of remote operation.
- It has a simultaneous measurement support with full navigation capability for GPS support and map displays of collected measurement data.
- It has utilities for data formatting and export.
- It supports many wireless technologies.

3.2.4 X-TEL

It is the first drive test tool to combine a real-time mapping interface and it is a first portable, fully expandable, modular platform for both indoor & outdoor data collection. It supports many wireless technologies. Its simultaneous data collection and analysis configuration is depicted in figure 3.4. It is designed to support all major digital phone standards and technologies. It offers more scanning devices, more phone technologies, and more data collection bandwidth than other drive test tools in the market. It has a flexible platform so it provides the ability to add ITU PESQ MOS scoring and remote control of drive test units using the most advance autonomous system on the market – Xi Autonomous.

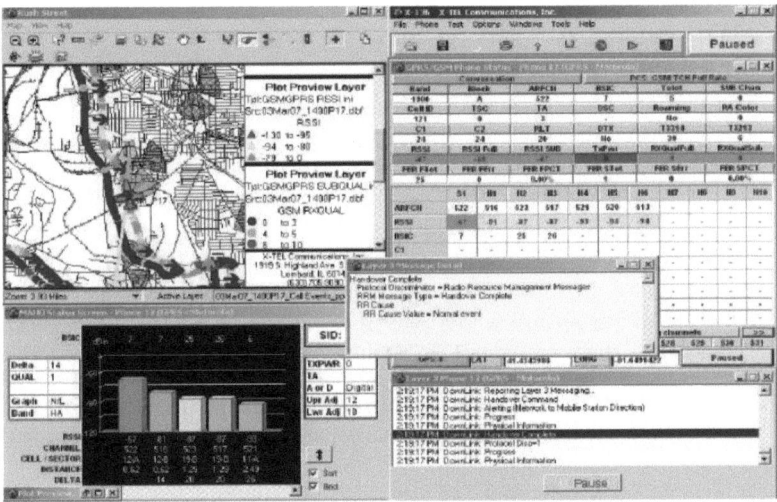

Figure 3.4 Window of XTEL's Data collection Tool

3.2.5 Test Mobile Stations (TEMS)

TEMS Investigation presents users tool with the ability to accumulate, evaluate, and post process the network data used on a daily basis for network checking, troubleshooting, and optimization. The complete tool kit of TEMS eliminates the need of multiple tools, reduces cost, time and effort of operations staff. In addition, TEMS Investigation gives users a way to utilize all the benefits of network data while at the same time protecting its integrity. TEMS Investigation collects data beyond the abilities of other tools in the market. It employs specially developed algorithms to collect exclusive information not offered in other vendor's tools. TEMS Investigation supports multimode functionality for system verification, troubleshooting, and optimization of WCDMA/HSPA, CDMA, and GSM/GPRS/EDGE networks. The multi-mode functionality of TEMS makes it possible to:

- Optimize intersystem handoff and cell selection.
- Check compressed mode behavior.
- Evaluate coverage and performance of base stations.
- Substantiate system accessibility of various wireless technologies like WCDMA/HSPA and GSM

TEMS Investigation is the test tool which is used for measurement, in current research and it is discussed in the following section

3.3 INTRODUCTION TO TEMS INVESTIGATION

TEMS Investigation is an air interface test tool for cellular networks and supports the following technologies [see app. III]:

- GSM/GPRS/EGPRS.
- WCDMA/HSDPA/HSUPA/HSPA+.
- TD-SCDMA.
- CDMA One/CDMA2000/1xEV-DO, plus basic iDEN support.

The product is also capable of scanning Long term evolution (LTE) and Worldwide Interoperability for Microwave Access (WiMAX) carriers. TEMS Investigation allows monitoring of voice and video telephony as well as various data services over packet switched and circuit switched connections. TEMS Investigation combines data collection, real time analysis and post processing all in one product. It is divided into two modules.

Module1: Data Collection

It is the part of TEMS Investigation that interfaces phones and other measurement devices, collects data, and records it in log files. It also allows presentation and analysis of a single log file at a time. It also includes major functionality found in TEMS Investigation CDMA 4.0.

Module2: Route Analysis

It is a module that permits rapid analysis of multiple log files, originating from TEMS Investigation itself or from TEMS Automatic, TEMS Drive Tester or TEMS Pocket. Statistical binning of log file data by area, time or distance is supported. The means of presentation – maps, line charts, and so on are fundamentally the same in both modules. Route Analysis also includes RAN Tuning which is a reporting tool for UMTS data (packet and circuit switching). RAN Tuning evaluates the network in terms of accessibility, mobility, coverage and retains ability.

(1) High Speed Packet Access (HSPA) is an mixture of two mobile telephony protocols, High Speed Downlink Packet Access (HSDPA) and High Speed Uplink Packet Access (HSUPA) that improves the performance of existing WCDMA protocols.

(2) Integrated Digital Enhanced Network (IDEN) is a popular wireless technology, developed by Motorola. It provides many benefits to its users like trunked radio and a cellular telephone. IDEN accompanies more users in a given spectral width, compared to analog cellular and two-way radio systems. it uses speech compression and time division multiple access (TDMA) techniques [146].

(3) Network Intrusion Detection System (NIDS) is an intrusion detection system that detects malicious activity like denial of service attacks, port scans or even attempts to crack into computers by monitoring network traffic.

(4) High Speed Uplink Packet Access (HSUPA) is a 3G mobile telephony protocol in the HSPA family with uplink speed is up to 5.76 Mbps. It was produced by Nokia. The official 3GPP name for 'HSUPA' is Enhanced Uplink (EUL).

(5) 3GPP Long Term Evolution (LTE) is the latest standard in the mobile network technology tree. It is a 3rd Generation Partnership Project (3GPP), operating under a name trademarked by the European Telecommunications Standards Institute. The current generation of mobile telecommunication networks is collectively known as 3G (for "third generation"). It is marketed as 4G. Its first release does not satisfy with the IMT Advanced 4G requirements. The pre-4G standard is a step toward LTE Advanced, a 4th generation standard (4G) of radio technologies designed to increase the capacity and speed of mobile telephone networks. LTE Advanced is backwards compatible with LTE and uses the identical frequency bands, while LTE is not backwards compatible with 3G systems.

(6) WiMAX (Worldwide Interoperability for Microwave Access) is telecommunications protocol that provides fixed and mobile internet access.

Some of the GSM parameters are given below which must be known in order to perform measurements [57].

3.4 GSM PARAMETERS AND THEIR RANGE

3.4.1 Handoff

In cellular telecommunications, When a mobile unit is moving from one core network to another network while a call is in progress, the mobile switching centre

(MSC) automatically transfers the call to a new network belonging to the new base station [10], [181]. Mobility management addresses two main problems those are location management and handoff management. Location management tracks the Mobile Terminals (MT) for loss less information delivery. Handoff management maintains the active connections for roaming mobile terminals as they change their point of attachment to the network [6].

There are different categories of GSM handoff which involves different parts of the GSM network. Changing cells within the same BTS is not complicated as the changing of the cell belonging to different MSC. There are mainly two reasons for this kind of handoff. The mobile station is moving out of the coverage range of base station or the antenna of BTS. Secondly the MSC or the BSC may decide that the traffic in the present cell is too high and transfer some traffic to another cell with lower traffic load. Different kinds of handoff are presented in the following section [176], [213], and [215]:

(a) Intra-cell BTS Handoff: The terms intra-cell and intra BTS handoff are used for frequency change. There is a slight difference between them but almost they are considered the same. The term intra-cell handoff is not real as it deals with the frequency change of a going call. The frequency changes occur when the quality of the communication link degrading and the measurements of the neighboring cells are better comparing with the current cell. The communication link degradation is caused by the interference as the neighboring cell using the same frequencies and it's better to try another channel. In the intra BTS handoff cell involved are synchronized.

(b) Intra-BSC Handoff: It is performed when the MSC changes the BTS instead of BSC. The intra-BSC handover is entirely carried out by the BSC, but the MSC is notified when the handoff has taken place. In the intra-BSC handoff both synchronized and non synchronized handoff are possible.

(c) Intra-MSC Handoff: In the intra-MSC handoff when the BSC decides that handoff is required but the targeted cell is controlled by different BSC then it needs assistance from the connected MSC. In comparison to the pervious handoff discussed the MSC mandatory for this kind of handoff. Responsibilities of the MSC do not include processing the measurements of the BTS or MSC but to conclude the handoff. This kind of handoff can be other intra-MSC or Inter-MSC. In the intra-MSC handoff the targeted cell is allocate in different BSC connected by the same MSC. The MSC contacts the targeted BSC for allocation of the required resources and inform the BSC when they are ready. After the successful resources allocation the MSC instructed to access the new channel and the call is transferred to the new BSC.

(d) Inter-MSC Handoff: The inter-MSC handoff is performed when the two cells belonging to different MSC in the same system. In the inter-MSC handoff the targeted cell is connected is connected to different MSC than the one currently serving the call MSC.

In telecommunications there are various reasons why a handoff (handover) might be conducted. Some of those are:

- When a mobile phone user is travelling from the area covered by one base station to the area covered by another base station then the call is transferred to the second base station to avoid call termination.
- When the capacity for connecting new calls of a given cell is used up and an existing or new call from a phone, which is located in an area overlapped by another cell, is transferred to that cell in order to free up some capacity in the first cell for other users, who can only be connected to that cell [73].
- In non-CDMA networks when the channel used by the phone is interfered with another phone using the same channel in a different cell, then the call is transferred to the second base station to avoid interference.
- In CDMA networks a soft handoff may be induced in order to reduce the interference to a smaller neighboring cell due to the "near-far" effect even when the phone still has an excellent connection to its current cell [123].
- Ping-Pong Handover effect is caused by fading. If the MS is moving in a zigzag pattern between the cells, or by non linearity in the receiver then this type of effect occur.

3.4.2 Rx Level

This parameter of GSM network is used in current research work. Drive test results will give the level of signals in different regions of the network. In urban areas, coverage is generally found to be less at the farthest parts of the network, in the areas behind high buildings and inside buildings. These issues become serious when important areas and buildings are not having the desired level of signal even when care has been taken during the network planning phase. The problems are usually solved out by changing the antenna locations and by altering the tilt of the antennas [93]. Coverage also becomes a major issue in rural areas, where the capacity of the cell sites is low. A factor that may lower the signal level could be propagation conditions, so study of link budget calculations along with the terrain profile becomes a critical part of the rural optimization [216]. For highway coverage, addition of new sites may be one of the solutions.

3.4.3 Rx Quality

The quality of the radio network is dependent on its coverage, capacity and frequency allocation. Most of the severe problems in a radio network can be attributed to signal interference. For uplink quality, BER statistics are used, and for downlink FER statistics are used [193]. The problems of interference may be caused due to flaws in the frequency plan, in the configuration plans (e.g. antenna tilts), inaccurate correction factors used in propagation models etc.

Table 3.1 Range of SQI

SQI values	Perceived speech quality
20 ≤ SQI ≤ 21 / 30	Very good for FR / EFR
1 ≤ SQI ≤ 19	good
SQI ≤ 0	bad

3.4.4 Speech Quality Index (SQI)

SQI is an estimate of the perceived speech quality as experienced by the mobile user, is based on handover events and on the bit error and frame erasure distributions. The SQI scale goes from -15 to 21 for Full Rate (FR) speech coders and from -15 to 30 for Enhanced Full Rate (EFR) speech coders. The table 3.1 shows the relation between SQI and perceived speech quality. From this table it has been observed that if the SQI values were less than or equal to 0 then the perceived speech quality is bad [172].

The field measurement is the basic need for this research work. The requirements for field measurements are mentioned below.

3.5 REQUIREMENTS FOR FIELD MEASUREMENT

The field measurement is the primary need for this research work. For collection of the field data in GSM frequency band, the hardware as well as the software is required.

3.5.1 Hardware Requirement

The following are the main hardware which required in field data measurement [9], [105]:

- Test Kit --- for hardware holding & protection
- Mobile Computer --- for test control & data collection
- Trace Mobile --- for wireless parameter collection
- GPS --- for geographic positioning
- Power Supply System --- Supplying power for GPS & notebook computer
- PCMCIA Card --- for communication between the notebook computer, GPS and
- Trace phone

3.5.2 Software Requirement

TEMS Investigation is network drive test software. It is combined with the above hardware to perform a GSM network test. TEMS provides with real time network parameters such as LAC, CI, Rx level, Rx quality etc on basis of a digital geographic map & cell database. The test drive principle for data measurement is shown in figure 3.5.

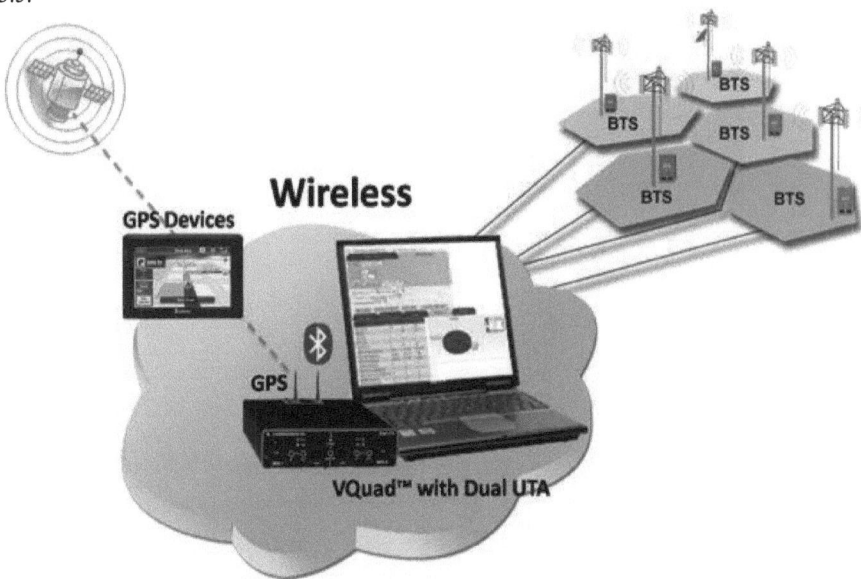

Figure 3.5 Test Principle Illustrations

3.5.3 Specification of Hardware and Software

Usually, the TEMS test system requires a hardware configuration in the following aspects [139]:

(A) TRACE PHONE: The following phones (supporting both GSM and WCDMA Unless otherwise stated) can be delivered with TEMS Investigation 10.0.

➢ Sony Ericsson C905, Sony Ericsson C905a, Sony Ericsson Z750i, Nokia N96 EU, Nokia N96 US (table 3.3 shows the features of these mobile phone)

  ➢ Other Connectable UMTS Devices

Sony Ericsson C702, Sony Ericsson K600i, Sony Ericsson TM506, Sony Ericsson W760i, LG CU320, LG CU500, LG U960, Motorola E1000, Motorola E1070, Motorola M2501, Motorola Razr Maxx V6, Motorola Razr V3x, Motorola Razr V3xx, Motorola Razr2 V9, Nokia 6120, Nokia 6121, Nokia 6280, Nokia 6680, Nokia 7376, Nokia N75, Nokia N80, Nokia N95

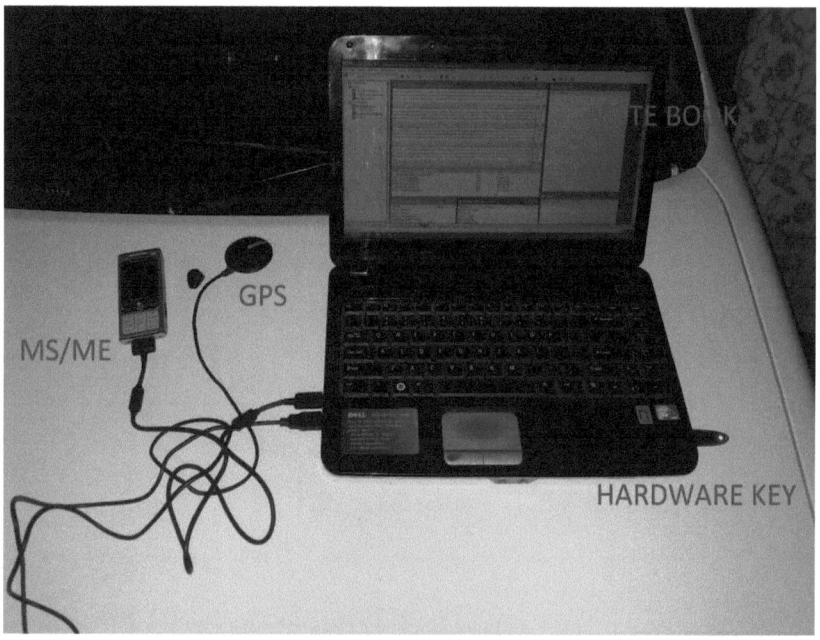

Figure 3.6 TEMS Test Kit used for test drive

It is recommended that a status configuration of the trace phone required as below prior to a test:

- Data Rate: 9600bps; Menu: Accessories\Data parameters\Speed
- Modem type: Data; Menu: Accessories\Data parameters\Mode
- Protocol: Secure link; Menu: Accessories\Data parameters\Protocol.

(B) GPS UNITS: TEMS Investigation supports the NMEA-0183 GPS protocol. Some other GPS units using a different protocol are also compatible with TEMS Investigation, including the GPS built into the supported PCTEL scanners. This is the full list of recommended GPS units [193], [160]:

- Bosch/Blaupunkt Travel Pilot
- Garmin 10 Mobile Bluetooth (NMEA-0183)
- Garmin 12XL (NMEA-0183)
- Garmin 35 (NMEA-0183)
- Global Sat BT-359 (NMEA-0183)
- Global Sat BT-368 (NMEA-0183)
- Global Sat BU-303 (NMEA-0183)
- GlobalS at BU-353 (NMEA-0183)
- HoluxGPSlim 236 (NMEA-HS; Bluetooth or USB)
- MagnettiMarelliRoutePlanner NAV200
- Nokia LD-3W Bluetooth (NMEA-0183)
- Sanav GM-44 (NMEA-0183)
- Sanav GM-158 (NMEA-0183)
- NMEA-HS HoluxGPSlim 236 BT
- NMEA-HS HoluxGPSlim 236 USB

Table 3.2 Some TEMS Supported Mobile Phone's Feature

Phone name /Feature	WCDMA 1900	WCDMA 2100	GSM 900	GSM 1800	GSM 1900	Video calling	External antenna
Sony Ericsson K790a	No	No	No	Yes	Yes	no	Yes
Sony Ericsson K790i	No	No	Yes	Yes	Yes	No	Yes
Sony Ericsson K800i	No	Yes	Yes	Yes	Yes	Yes	Yes
Motorola Razr V3xx Eu	No	Yes	Yes	Yes	Yes	Yes	No
Motorola Razr V3xx US	Yes	No	No	Yes	Yes	No	No
Motorola Razr2 V9 EU	No	Yes	Yes	Yes	Yes	Yes	No
Motorola Razr2 V9 US	Yes	No	No	Yes	Yes	No	No
NOKIA 6096	No	No	Yes	Yes	Yes	No	No
NOKIA 6125	No	No	Yes	Yes	Yes	No	No
NOKIA N75	yes	No	Yes	Yes	Yes	No	No
NOKIAN80	No	Yes	Yes	Yes	Yes	Yes	No

For poor receiving conditions one of the products Bosch/Blaupunkt Travel Pilot or Magnetti Marelli Route Planner can be recommended, since they have dead reckoning facilities. The GPS unit which has been used in data collection is shown in figure 3.6.

> (C) NOTEBOOK: A notebook computer is used to run the test software TEMS Investigation 10.0 record the test data meanwhile monitor the test status. In order to ensure a perfect test performance, a basic configuration of the notebook illustrated below is recommended when running the TEMS for GSM network test.
>
> - CPU: PII 400MHz
> - Memory: 128M SDRAM
> - Hard Disk Drive: 10GB
> - Display Resolution: 1024 * 768 pixel
> - Operation System

The following operating systems are supported:

- Windows Vista with Service Pack 1, Enterprise Edition
- Windows XP Professional with Service Pack 3
- Windows XP Tablet PC Edition with Service Pack 3

In addition, all the latest Windows updates should always be installed. Supported languages are English (U.S.), Chinese (simplified characters), and Japanese. The note book used for field data collection is shown in figure 3.6.

> (D) CABLES TO USE WITH PHONE: Sony Ericsson C702, C905, C905a, TM506, W760i, Z750i. It is necessary to use the USB cable supplied with the phone (DCU-65, part number RPM 131 12/1 R1B). Using a USB cable delivered with older phones such as K800i and K790i is not recommended.
>
> (E) EXTERNAL ANTENNA TO PHONES: An external monopole antenna is connected to a phone by means of a coaxial cable (and possibly an adapter) supplied with the phone. Phone specific details are as follows:

- Sony Ericsson Z750i

An external antenna with the Sony Ericsson Z750i, the phone is delivered with the antenna adapter already mounted. Just connect the antenna to the adapter. A Sony Ericsson Z750i phone with an external antenna mounted cannot use its internal antenna.

- Nokia N96

Press the antenna adapter into the hole provided for it, and then connects the antenna to the adapter.

(F) PCMCIA CARD: A PCMCIA card is used to connect the trace phone and GPS to a notebook computer. The test data will be transferred from the trace phone and GPS to the notebook computer through such a connection.

(G) POWER SUPPLY SYSTEM: This system supplies power for both the notebook and GPS. The power source comes from a car supply by linking a power cable with the car ignition during the test. The power cable & universal adapter (for notebook power supply) are contained in it.

3.6 ASSEMBLING / INSTALLATION / SETUP PROCEDURE

The assembling / setup procedure plays very important role. Setup procedure for drive test is presented in the following section.

3.6.1 Plugging Phones & Data Cards

Nearly all supported phones connect to the laptop via USB. A single USB cable connects the USB port on the phone to a USB port on the laptop; both TEMS measurements and data service measurements are transferred through this cable [26].

3.6.2. Plugging in GPS Units

Supported GPS units connect via USB or Bluetooth, or to a COM port.

3.6.3. Configuring TEMS Investigation for Data Collection

Launch the application from the Start menu [98]:

Choose Start → Programs → Ascom → TEMS Products → TEMS Investigation 10.0 Data Collection.

In Windows, it must run the application as administrator. This option is selected by right clicking the Start menu item above and choosing Properties as shown in figure 3.7.

→ Shortcut tab → Advanced

All devices that are known to TEMS Investigation, whether automatically detected or manually added, are listed in the Equipment Configuration window. In the example below, a Sony Ericsson C702 phone has been auto detected [98], [128]. In the Name column of the Equipment Configuration window, each device is represented by an EQ item, containing within it one further item (called "channel") for each data source furnished by the device:

➢ for user terminals, an "MS" channel representing TEMS measurements and a "DC" channel representing data service measurements
➢ for scanners, an "MS" channel
➢ for GPS units, a "PS" channel (P for "positioning"). GPS units built into scanners also appear as separate devices.

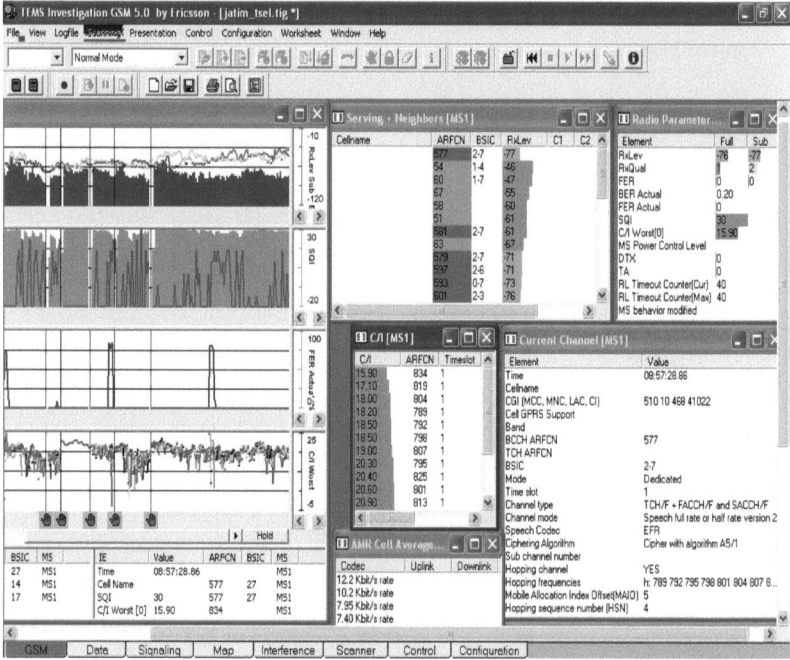

Figure 3.7 TEMS window

The colored dot to the left of each channel shows the status of the channel in TEMS Investigation: red means "not connected", green means "connected".

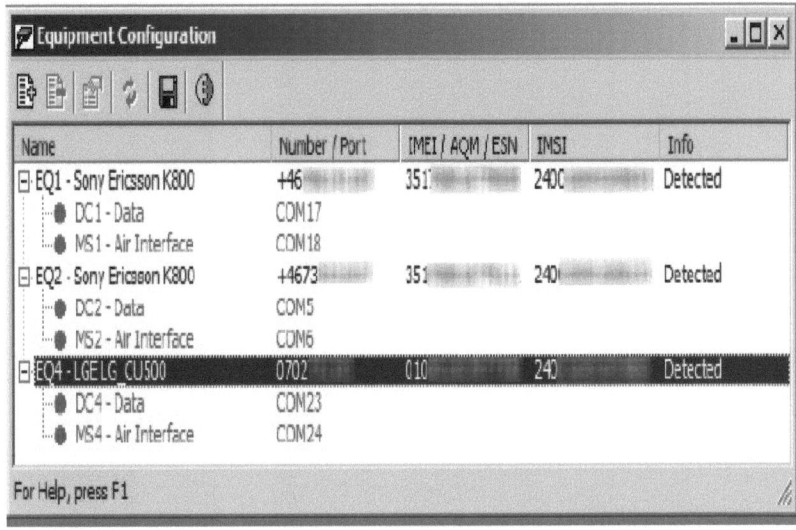

Figure 3.8 Equipment Configuration Window in TEMS

Saving of the Equipment Configuration is given in the figures 3.8 and 3.9.

You can save the current setup in the Equipment Configuration window:

Click the Save button

Connecting External Equipment

Connecting external devices in the application is done as a separate step. To connect a single channel of an external device:

• Select the channel in the combo box of the Equipment Control toolbar.

Click Connect on the Equipment Control toolbar
To connect all channels of all external devices:

Click Connect all on the Connections toolbar.
Connected channels are accompanied by a green light symbol in the combo box. The symbols on the status bar and in the Equipment Configuration window likewise turn green.

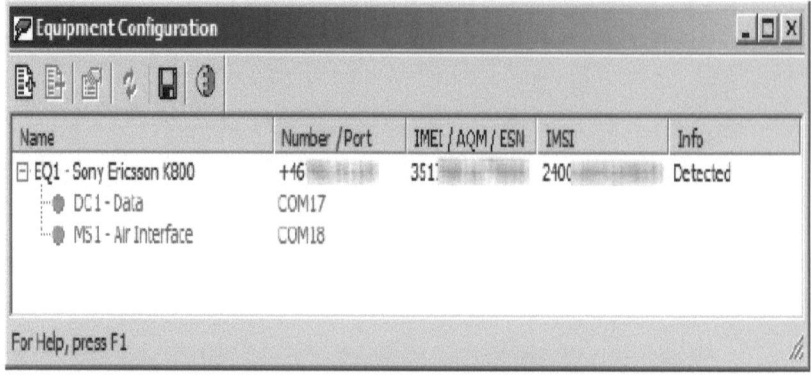

Figure 3.9 Equipment Configuration Window

Recording Log files

Log files can be recorded in the following ways in TEMS Investigation:

• From the Record toolbar or Log file menu
• From within command sequences.

All these procedures produce the same type of output file, with extension log files. To initiate recording of a log file:

Click Start Recording on the Record toolbar.
• Perform the desired tasks with your external equipment.

To pause the recording without closing the log file, click Pause Recording.

Click the same button once more to resume the recording. Events indicating pausing and resumption are written to the log file.

Click Stop Recording to end the recording and close the log file. Once you have closed it, you cannot log any more data to the same file.

Loading Site data, Vector, Map, Path Cell Data

TEMS Investigation can present information on individual cells in cellular networks. In particular, it is possible to draw cells on maps and to display cell names in various windows as shown in figure 3.10. Cell data is also made use of in log file reports. Cell data can be provided in two ways:

- In a plain text XML file (*.xml) whose format is common to several TEMS products.
- In a file with a plain text, TEMS Investigation specific format (*.cel). This format is for UMTS only. GSM and WCDMA cells can be mixed in one file.

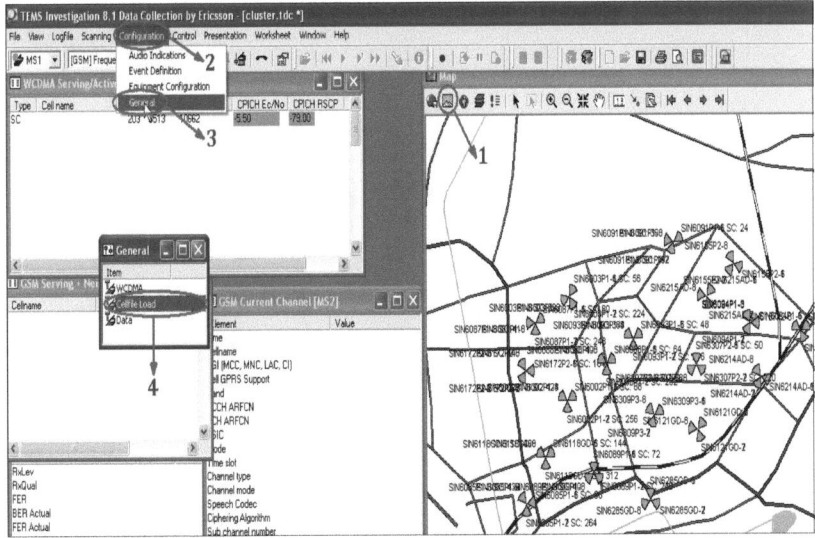

Figure 3.10 Cell Data Configuring Window

Loading Cell Files

To make a cell file active, it must be loaded in the General window and several cell files loaded in the application at the same time is possible here.

➢ From the Navigator, open the General window.
➢ In the General window, double-click the item "Cell file Load".
➢ To add a cell file, click the Add button and browse to select your file. The cell file is added in the list box.
➢ To remove a cell file from the list, select it and click Remove. To remove all cell files, click Remove all.
➢ When you are done selecting cell files to load, click OK.

The set of loaded cell files can be modified at any time. If multiple files of the same type (CEL or XML) are loaded, the information in all files is correlated in the presentation. However, if you load both CEL and XML files, no attempt is made to correlate CEL and XML cell information; rather, cell information is presented separately from each type of file. To load a map file, right-click the Map Files item in the Navigator and choose Add from the context menu, then browse for the desired TAB or GST file. The loading of cell files with drive test results may be understand in better way with the help of figure 3.11.

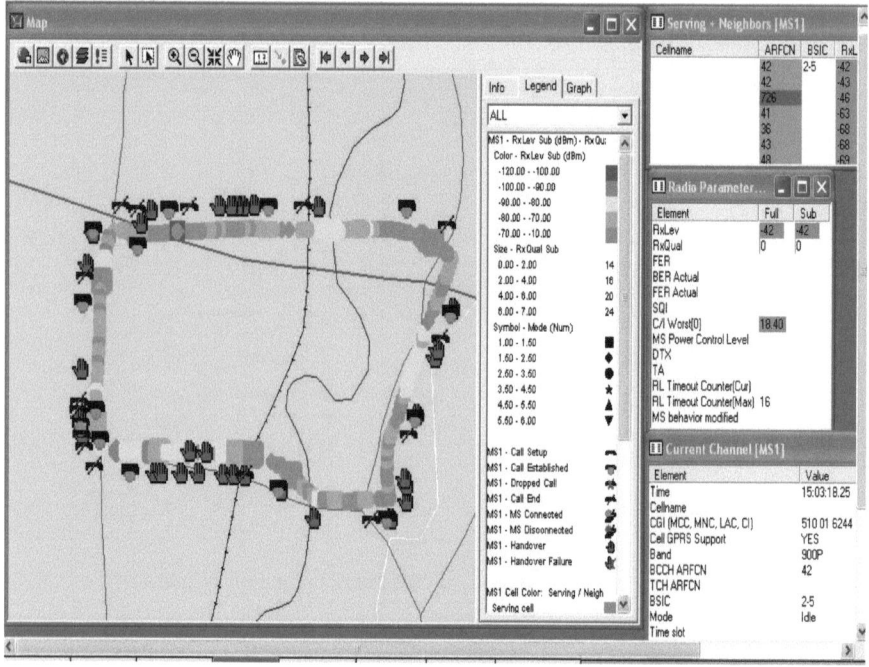

Figure 3.11 Example of loading of Cell File in Narnaul (South Haryana)

3.6.4. Test Procedure

After connecting the drive test tool, following activities has been carried out during Drive test:

- TEMS Investigation application software is opened in the Laptop. The system by default opens 'GSM' window displaying the empty tables and charts meant for RF information.
- Both external devices, TEMS mobile handset and GPS has been configured in the 'Control and configuration' window. The system detects the devices but is indicated as disconnected with the red color symbol. This changes to green color after clicking 'Connect All' in the Connection Toolbar.
- The mobile is then connected in the 'idle mode'. The GSM window starts displaying the live network data in the corresponding tables and charts. GPS window shows Latitude & Longitude of the place.

- ➢ 'Record' tool bar has been initiated. Log file has been saved followed by originating call on the TEMS phone. The test enters in the 'dedicated mode'.
- ➢ Driving has been carried out on routes which cover the cell and all neighboring cells.
- ➢ Cell coverage, Received signal strength, Quality and many other RF parameters has been collected. Call connection, call mobility control, call release and many other events has been checked and recorded.
- ➢ Drive test log file is generated and exported to different formats for analysis,

In a field measurement both radial and azimuthally routes has been tested. In urban area the effect of street orientations has been considered. Both Line of Site (LOS) and non-LOS points have to be included in the drive test. Measurements is unbiased i.e. the data collected should represent all typical coverage scenarios [10], [109]. The selection of drive test routes is based on following parameters:

- ➢ The terrain variations.
- ➢ Subscriber distributions.
- ➢ Major highways and thoroughfares.
- ➢ Critical Areas.
- ➢ Potential Shadowing areas and handover regions.

Distance from site to be driven depends on the cell size and typically 2-3 times the cell radius is drive tested. Current analysis is limited to receive signal strength and path loss calculations. Five GSM base stations has selected for measurement in Narnaul (Haryana, India). Mainly the cell ids NNL001, NNL002, NNL003 have been used for field data collection because this research work is concentrated on urban area. The field data collected during test drive in these three different cell ids has been compared with different path loss prediction models. Coordinates of these base stations are given below:-

- ➢ Base station (ID NNL001) - Lon. 76.1105 , Lat. 28.0413 ⎫
- ➢ Base station (ID NNL002) - Lon. 76.1125 , Lat. 28.0427 ⎬ — Urban Area
- ➢ Base station (ID NNL003) - Lon. 76.1145 , Lat. 28.0445 ⎭
- ➢ Base station (ID NNL004) - Lon.76.1154 , Lat. 28.0459   Sub urban Area
- ➢ Base station (ID NNL005) - Lon. 76.1178, Lat. 28.0495   Rural Area

The results of these three selected base stations in terms of the received signal strength are presented in following tables:

Table 3.3 Signal Strength Measurements at Base Station NNL001 in Month of January (Winter)

S. No	Distance from base station (km)	Received signal strength (dBm)
1	0.1	-59
2	0.2	-66
3	0.3	-70
4	0.4	-65
5	0.5	-69
6	0.6	-67
7	0.7	-71
8	0.8	-73
9	0.9	-76
10	1.0	-75
11	1.1	-78
12	1.2	-79
13	1.3	-75
14	1.4	-75
15	1.5	-80
16	1.6	-78
17	1.7	-81
18	1.8	-79
19	1.9	-81
20	2.0	-84

Table 3.4 Signal Strength Measurements at Base Station NNL011 in Month of may (Summer Temperature:47^0 C)

S. No	Distance from base station (km)	Received signal strength (dBm)
1	0.1	-58
2	0.2	- 63
3	0.3	-66
4	0.4	-71
5	0.5	-66
6	0.6	-72
7	0.7	-74
8	0.8	-75
9	0.9	-76
10	1.0	-74
11	1.1	-74
12	1.2	-77
13	1.3	-79
14	1.4	-75
15	1.5	-77
16	1.6	-79
17	1.7	-83
18	1.8	-78
19	1.9	-81
20	2.0	-85

Table 3.5 Signal Strength Measurements at Base Station NNL001 in Month of July (Heavvy rain)

S. No	Distance from base station (km)	Received signal strength (dBm)
1	0.1	-56
2	0.2	- 62
3	0.3	-63
4	0.4	-67
5	0.5	-71
6	0.6	-75
7	0.7	-77
8	0.8	-74
9	0.9	-71
10	1.0	-80
11	1.1	-74
12	1.2	-81
13	1.3	-78
14	1.4	-78
15	1.5	-79
16	1.6	-82
17	1.7	-83
18	1.8	-82
19	1.9	-84
20	2.0	-86

Table 3.6 Signal Strength Measurements at Base Station NNL001 in Month of December (Winter heavy fog condition)

S. No	Distance from base station (km)	Received signal strength (dBm)
1	0.1	-60
2	0.2	-65
3	0.3	-68
4	0.4	-73
5	0.5	-76
6	0.6	-71
7	0.7	-76
8	0.8	-72
9	0.9	-74
10	1.0	-80
11	1.1	-75
12	1.2	-79
13	1.3	-77
14	1.4	-80
15	1.5	-83
16	1.6	-79
17	1.7	-84
18	1.8	-86
19	1.9	-83
20	2.0	- 91

Table 3.7 Signal Strength Measurements at Base Station Hisar in Month of January

S. No	Distance from base station (km)	Received signal strength (dBm)
1	0.1	-59
2	0.2	- 66
3	0.3	-70
4	0.4	65
5	0.5	-69
6	0.6	-67
7	0.7	-71
8	0.8	-73
9	0.9	-76
10	1.0	-75
11	1.1	-78
12	1.2	-79
13	1.3	-75
14	1.4	-75
15	1.5	-80
16	1.6	-78
17	1.7	-81
18	1.8	-79
19	1.9	-81
20	2.0	-84

Table 3.8 Signal Strength Measurements at Base Station Hisar in Month of May

S. No	Distance from base station (km)	Received signal strength (dBm)
1	0.1	-58
2	0.2	- 63
3	0.3	-66
4	0.4	-71
5	0.5	-66
6	0.6	-72
7	0.7	-74
8	0.8	-75
9	0.9	-76
10	1.0	-74
11	1.1	-74
12	1.2	-77
13	1.3	-79
14	1.4	-75
15	1.5	-77
16	1.6	-79
17	1.7	-83
18	1.8	-78
19	1.9	-81
20	2.0	-85

Table 3.9 Signal Strength Measurements at Base Station Hisar in Month of july

S. No	Distance from base station (km)	Received signal strength (dBm)
1	0.1	-56
2	0.2	-62
3	0.3	-63
4	0.4	-67
5	0.5	-71
6	0.6	-75
7	0.7	-77
8	0.8	-74
9	0.9	-71
10	1.0	-80
11	1.1	-74
12	1.2	-81
13	1.3	-78
14	1.4	-78
15	1.5	-79
16	1.6	-82
17	1.7	-83
18	1.8	-82
19	1.9	-84
20	2.0	-86

Table 3.10 Signal Strength Measurements at Base Station Hisar in Month of December

S. No	Distance from base station (km)	Received signal strength (dBm)
1	0.1	-60
2	0.2	-65
3	0.3	-68
4	0.4	-73
5	0.5	-76
6	0.6	-71
7	0.7	-76
8	0.8	-72
9	0.9	-74
10	1.0	-80
11	1.1	-75
12	1.2	-79
13	1.3	-77
14	1.4	-80
15	1.5	-83
16	1.6	-79
17	1.7	-84
18	1.8	-86
19	1.9	-83
20	2.0	-91

3.7. VARIOUS ISSUES DURING MEASUREMENT

While on field measurements and testing, some issues similar to other researchers [15], [236] were raised are discussed here.

3.7.1. Overshooting

Any site sector x overshooting indicates that it serves at very long distance with good Rx Level in the particular area and it is supposed to be served by another closer site having nearly the same signal strength. This may cause high interference leading to bad quality. Usually the recommendation for this case is to down tilt the antenna.

3.7.2. Bad Quality

Bad quality occurs due to bad coverage and high interference. The bad coverage can be rectified by checking the infected site or antenna. The high interference can be rectified by rechecking the RF planning using planning tool.

3.7.3. Bad Coverage

When people are standing within the main beam of the antenna at a reasonable distance and still the coverage is weak. This case occurs due to wrong azimuth or low power transmission from the BTS or due to hardware failure.

3.7.4. Missing Neighbors

This case may cause to drop calls and bad coverage as the MS will not be able to handover on the target site as it's not known for the serving site.

3.7.5. Dropped Calls

A drop call may occur due to the above case as mentioned. It may also occur due to an uncharacteristic release in the TCH or SDCCH after it's being occupied successfully. In the downlink case the radio link time out may decrease to zero.

3.7.6. Blocked Calls

It's usually due to congestion on the serving cell or the assigned BCCH is highly interfered. Practical call connections is established at maximum distance from the site with timing advance of 68. If the site is near a source of water, for e.g. sea or a canal, signal interference occurs due to water reflection. Therefore if the MS is served by this site at this huge distance with a fair signal, block calls will occur.

3.7.7. Handover Failure & Delay

High handover failure will probably be due to:

- High neighbor interference.
- High source cell interference.
- Mobile on incorrect source cell.
- Mobile allocated parameter is incorrect or unexpected neighbor.
- Location area borders planned poorly.
    - Borders on road junctions or ridges.
    - Borders on large water expanses.
- Cell borders planned poorly.
    - Borders on road junctions or ridges.
    - Borders on large water expenses.
- Database parameters.
    - Power budget algorithms incorrectly specified.
    - Rx Quality algorithms incorrectly specified.
    - Rx Level algorithms incorrectly specified.
    - Timing advance algorithms incorrectly specified.
- No dominant server.
    - Neighbors being received at similar levels.
    - Missing cell site.
- The reason for handover delay is that some of the handover parameters (handover watch time, handover valid time) need to be adjusted.

3.8 DATA ANALYSIS AND SIMULATION TOOLS

Log files which are generated by TEMS Investigation can be analyzed on the following tools by exporting the log file into .tab format. The following are the software which can be used:

- MapInfo
- Arc View
- Marconi Planet DMS 3.1
- Ethereal

- MDM (CDMA)
- MATLAB

Here the MapInfo software has been used for generating log files from the collected field data and these data were simulated in MATLAB.

3.8.1 MapInfo

MapInfo Professional software was commenced in the year 1986. This was the first desktop GIS product for Mapping Display and Analysis System (MIDAS) [93]. MapInfo Professional is desktop Geographic information system or GIS software product created by Pitney Bowes Software, formerly MapInfo. GIS generally used for mapping and location analysis. Initially It was accessible only for the DOS operating system. This software might be modified by using the Map Code development environment. The first name of this software was Pitney Bowes Software and this name was subsequently changed to "MapInfo" after the second release. The DOS product was entirely replaced by MapInfo for Windows in the year 1990. MapInfo with graphical user interface (GUI) was prepared for the Microsoft Windows, UNIX and various operating systems. The language of Map Code was swapped with a new language Map Basic. These maps are organized into three layers as shown in figure 3.12. These layers resemble transparencies that are stacked on top of one another. Each layer holds different features of the entire map. In this software the user can begin by opening own table of data and displaying it in a Map window. In addition to this, the each layer contains style overrides and zoom layering characteristics [42].

These Map layers shape the basic building blocks of maps in MapInfo Professional. After the creation of map layers, user customizes them various ways. For example, one layer may contain state boundaries, a second layer may have symbols that represent capitals, a third layer might consist of text labels. By stacking these layers one on top of the other map layer, user can build a complete map. There are five basic objects of mapinfo as shown in figure 3.13.

3.8.2 MATLAB

Matrix laboratory (MATLAB) is a high performance language for technical computing. This programming language integrates computation, revelation and programming in an easy to use environment. Using MATLAB various problems and their solutions are expressed in recognizable mathematical notation [124], [126], [235]. It is an interactive language system in which basic data element is an array and allows us to solve many technical computing problems. MATLAB has many advantages compared to conventional computer language (C, FORTAN, JAVA etc.) for technical problem solving.

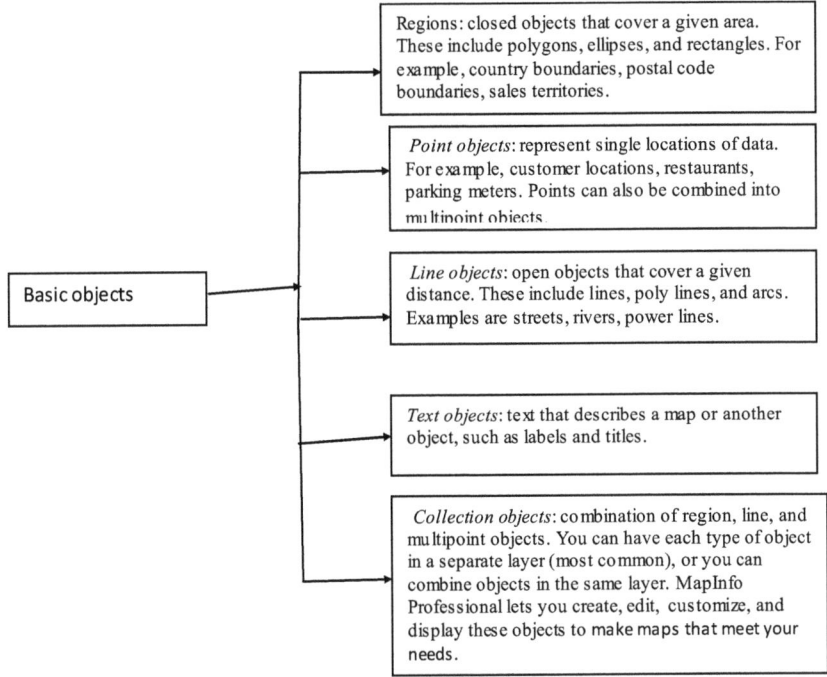

Figure 3.12 Five basic objects of Map info

- ➢ This software mainly uses matrices for internal operations i.e., the basic data element is a matrix.
- ➢ It is very easy to learn and easy to use programming language. It has fast prototyping for many programs.
- ➢ The functionality of the software can be expanded by adding various toolboxes to its environment according to the requirement. Tool box is a set of specific functions that provided more specialized functionality.
- ➢ This software supports various computer systems (platform independence). Program written in MATLAB can transfer to new platform according to the user requirements.
- ➢ It has some peculiar tools which allow a programmer to build a graphical user interface (GUI) for his program.
- ➢ It has a broad library of predefined functions which provide solutions to many basic technical operations.
- ➢ Flexibility and platform independence is achieved by compiling MATLAB programs into a device-independent p-code, and then interpreting the p-code instructions at runtime.

MATLAB has two principal disadvantages:

- As it is very heavy software, it occupies a huge amount of memory in the computer and slows down the speed of the computers.
- This software uses as much CPU time as Windows allows it to have and makes real-time applications very complicated.

A graphical user interface (GUI) is a graphical interface to a program. A fine GUI can construct programs easier to use by providing them with a reliable appearance and with instinctive controls like pushbuttons, list boxes, sliders, menu etc. GUI provides the user with a familiar environment in which to work. This GUI environment comprises pushbuttons, toggle buttons, lists, menus, text boxes, etc., MATLAB Systems of five main parts has been shown in table 3.11.

One can start MATLAB by double-clicking on the MATLAB icon. The main MATLAB window, called the MATLAB Desktop, the default desktop is shown in the figure 3.15.

Execution of MATLAB integrates many tools for running files, variables and applications within the MATLAB environment. The major tools within or accessible from the MATLAB desktop has been shown in figure 3.14 and table 3.12.

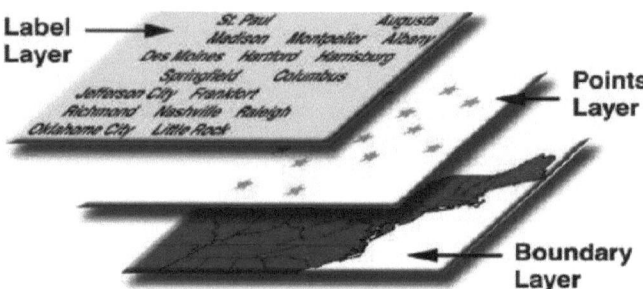

Figure 3.13 Illustration of Map Layer

Table 3.11 Main Parts of MATLAB

Main part	Application
MATLAB language	This is a high-level matrix/array language with control flow statements, functions, data structures, input/output, and object-oriented programming features. It allows both "programming in the small" to rapidly create quick and dirty throw-away programs, and "programming in the large" to create complete large and complex application programs.
Working environment	This is the set of tools and facilities that you work with as the MATLAB user or programmer. It includes facilities for managing the variables in workspace and importing and exporting data. It also includes tools for developing, managing, debugging, and profiling M-files.
Handle Graphics	This is the MATLAB graphics system. It includes high-level commands for two- dimensional and three-dimensional data visualization, image processing, animation, and presentation graphics. It also includes low-level commands that allow user to fully customize the appearance of graphics as well as to build complete Graphical User Interfaces.
Mathematical function library	This is a collection of computational algorithms ranging from elementary functions like sum, sine, cosine, and complex arithmetic, to more sophisticated functions like matrix inverse, matrix eigen values, Bessel functions, and fast Fourier transforms.
Application Program Interface (API).	This is a library that allows you to write C and Fortran programs that interact with MATLAB. It include facilities for calling routines from MATLAB (dynamic linking), calling MATLAB as a computational engine, and for reading and writing MAT-files.

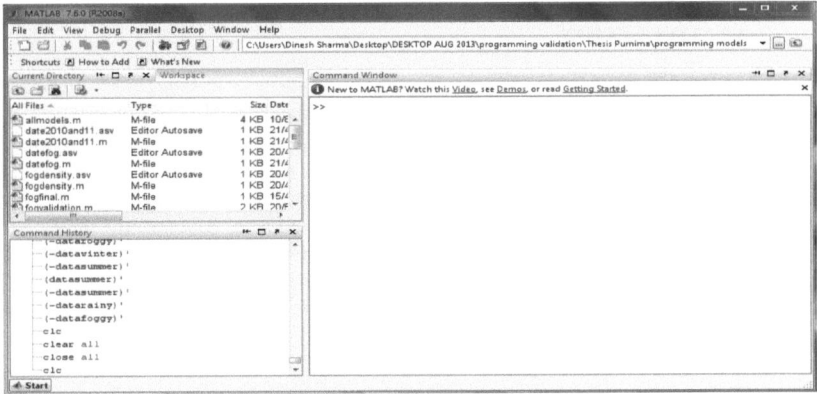

Figure 3.14 The Default MATLAB Desktop

Table 3.12 Different parts of MATLAB Window

MAJOR TOOL	APPLICATION
The Current Directory window	It shows a current directory with a listing of its contents. This window is useful for finding the location of particular files and scripts so that they can be edited, moved, renamed or deleted.
Workspace window	It provides a list of all the items in the workspace that are currently defined. These items consist of the set of arrays whose elements are variables or constants and which have been constructed or loaded during the current MATLAB session and have remained stored in memory.
Command Window	It locates in the default window where the command line prompt for interactive commands is located. Click in the command window to make it active. When a window becomes active, its title bar darkens. The ">>" is called the Command Prompt, and there will be a blinking cursor right after it waiting for you to type something. One can enter interactive commands at the command prompt (>>) and they will be executed on the spot.
Command History Window	It contains many commands that can be executed within the command window. This is a convenient feature for confirmation or checking of commands when debugging programs in a particular sequence during a multi-step calculation from the command line
Help Window	Typing help at the command prompt will reveal a long list of topics on which help is available. In addition, one can also get help on the certain command. Online help is also available in this window.

3.9 CONCLUSION

In this chapter the main focus was on different data collection and simulating software. The TEMS data collection tool has been used in present research work for field data collection. For field data collection the five cell ids has been selected. Out of these five cell, three are from urban area, one is from suburban and one is taken in rural area. But this research is in urban area, therefore the three cell ids has been taken for field data collection. During the test drive the field data has been collected in different climatic conditions. The log files were generated with the help of the MapInfo. After the field data collection, these data has been simulated in MATLAB for the comparative analysis.

CHAPTER 4

PERFORMANCE ANALYSIS OF DIFFERENT FIELD PROPAGATION MODELS

In this chapter, the analysis of different Empirical field propagation models proposed by earlier researchers has been carried out to compare their results with field data presently collected surrounding to south Haryana. The analysis of those propagation models have been made with the help of MATLAB. On the basis of this analysis a best fit propagation model i.e., Okumura model has been proposed for the south Haryana region.

4.1 INTRODUCTION

In mobile radio systems, propagation models (to predict radio propagation) are necessary for proper network planning process, interference estimations, frequency assignments, and estimation of cell parameters [11]. Most of the radio propagation models are deduced by combining analytical and empirical methods. Severe diffraction losses generally occur in areas, where there is no direct LOS path between the transmitter and receiver.

The accuracy of any propagation model is better realized when the input databases are exact and up to date. The propagation models are empirical mathematical formulae for the characterization of radio wave propagation as a function of frequency, distance and other parameters.

The analysis part of different field propagation models along with the currently collected field data from drive test at south Haryana region are described in the following section.

4.2 FIELD COLLECTED/ MEASURED DATA FROM DRIVE TEST AT SOUTH HARYANA REGION

Drive test is a common way to understand network performance by means of coverage evaluation, system availability, network capacity and quality of call. The data that has been taken during drive test gives idea only on the downlink side of the process. Investigation of various factors of the wireless network during drive test is given below:

- Working/Non-working sites
- Active/Inactive radio network features like frequency hopping
- Enabled/Disabled GPRS
- Overshooting sites – coverage overlaps/ Coverage holes
- Carrier to interference ratio analysis
- Drop Calls
- Capacity Problems
- Main Interference Sources/ High Interference Spots
- One-way/missing neighbors
- Handoff information
- Accessibility/ Retainability of the Network
- Equipment Performance / Faulty Installations

In drive test, Five GSM base stations are selected for measurements. Out of five cell id's three are from urban area, one is from suburban area and one is from rural area. The Coordinates of those selected base stations are given below:-

1. Base station (ID NNL001) - Lon. 76.1105 , Lat. 28.0413 ⎫
2. Base station (ID NNL002) - Lon. 76.1125 , Lat. 28.0427 ⎬ Urban Area
3. Base station (ID NNL003) - Lon. 76.1145 , Lat. 28.0445 ⎭
4. Base station (ID NNL004) - Lon.76.1154 , Lat. 28.0459   Sub urban Area
5. Base station (ID NNL005) - Lon. 76.1178, Lat. 28.0495   Rural Area

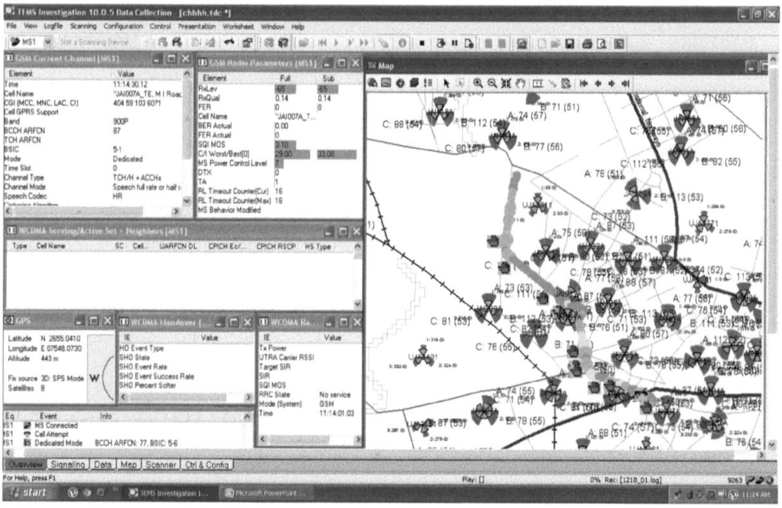

Figure 4.1 Selected Cell Sites for Field Data Collection

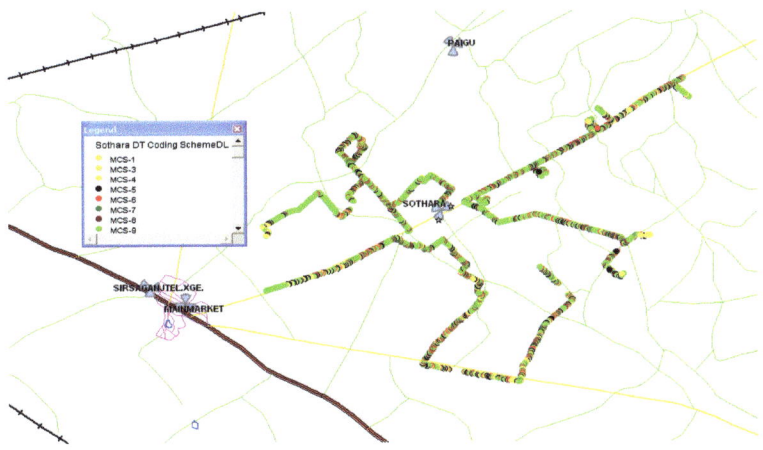

Figure 4.2 Drive Test Result in Cell id NNL001

4.3 RECEIVED SIGNAL STRENGTH AND PATH LOSS IN DIFFERENT AREAS

The Signal strength of field data collection depends on the distance between the transmitter and the receiver, environment and the area in which the drive test has been performed.

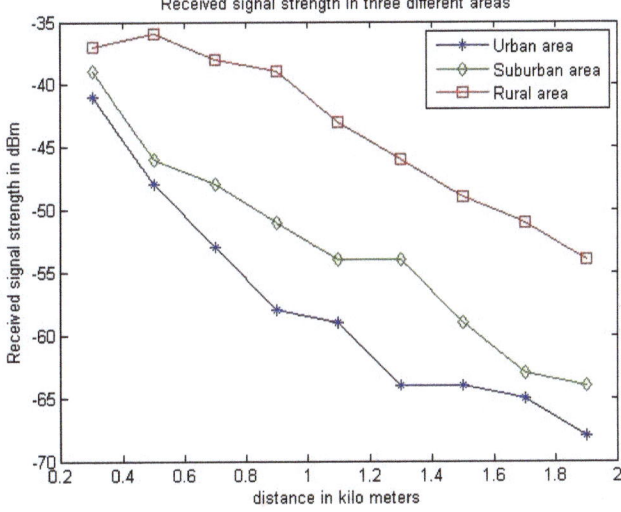

Figure 4.3 Variation of Received Signal Strength (dBm) with Distance (Km.) in Three Different Areas of Five Different Cell ids

The field data collection has been carried out at Narnaul (Haryana) in three different regions Rural, Urban and Suburban. The figure 4.3 depicts the variation of received signal strength with distance in different areas of Narnaul city and its nearby regions.

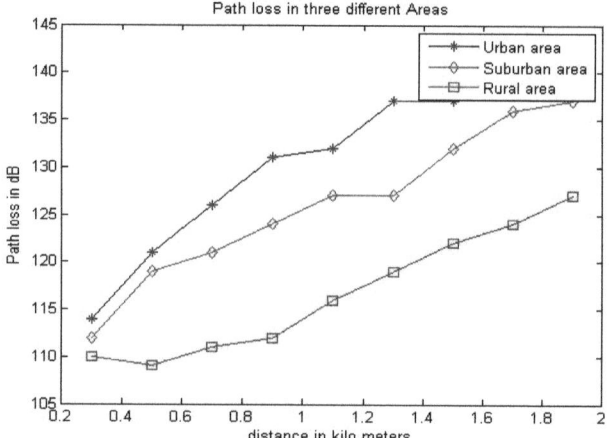

Figure 4.4 Variation of Path Loss (dB) with Distance (Km.) in Three Different Areas of Five Different Cell ids

The figure 4.3 and figure 4.4 represent the variation of signal strength and path loss during the drive test in three different areas. For the better understanding of the signal variation, the data collection is carried out in three different regions. But current work is concentrated on urban area of Narnaul city (South Haryana). It is found that the path loss in urban area is more as compared to the suburban and rural area. In other words in urban area, the received signal strength at a distance from the transmitter is less as compared to the suburban and rural area; this is due to huge number of buildings, environment and foliage etc exists in an urban area.

The analysis part of different field propagation models are described here.

4.4 COMPARISON BETWEEN FIELD MEASURED DATA AND PROPAGATION PATH LOSS MODELS IN SOUTH HARYANA

The field measured data like received signal strength, path loss and the distance between transmitter and receiver has been recorded in dBm and in Km unit. The field measurement has been taken from three base stations having ids NNL001, NNL002 and NNL003. The table 4.1 and 4.2 depict the average signal strength and average path loss in three different cells ids respectively.

Table 4.1 Average Signal Strength Measurements

Distance from base station (km)	Received signal strength (dBm) Site id:NNL001	Received signal strength (dBm) Site id:NNL002	Received signal strength (dBm) Site id:NNL003
0.1	-59	-58	-48
0.2	-71	-69	-55
0.3	-70	-71	-59
0.4	-65	-71	-66
0.5	-69	-65	-63
0.6	-67	-77	-67
0.7	-71	-74	-75
0.8	-73	-75	-77
0.9	-76	-76	-68
1.0	-80	-71	-73
1.1	-78	-68	-72
1.2	-79	-75	-76
1.3	-75	-66	-80
1.4	-68	-75	-83
1.5	-80	-77	-76
1.6	-74	-79	-74
1.7	-81	-83	-82
1.8	-83	-81	-74
1.9	-86	-89	-78
2.0	-88	-87	-82

Table 4.2 Average Path loss Measurements

Distance from base station (km)	Path loss(dB) Site id:NNL001	Path loss(dB) Site id:NNL002	Path loss(dB) Site id:NNL003
0.1	132	131	121
0.2	144	142	128
0.3	143	144	132
0.4	138	144	139
0.5	142	138	136
0.6	140	150	140
0.7	144	147	148
0.8	146	148	150
0.9	149	149	141
1.0	153	144	146
1.1	151	141	145
1.2	152	148	149
1.3	148	139	153
1.4	141	148	156
1.5	153	150	149
1.6	147	152	147
1.7	154	156	155
1.8	156	154	147
1.9	159	162	151
2.0	161	160	155

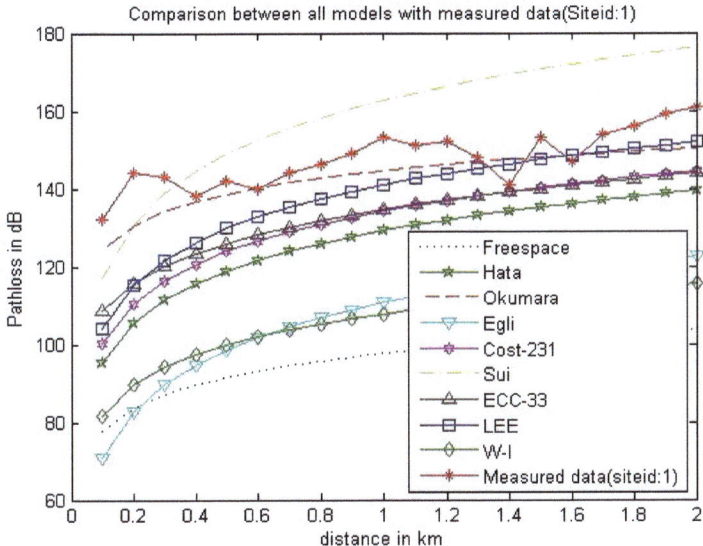

Figure 4.5 Comparison of Field Measured Path Loss and Predicted Path Loss with Distance (Site id NNL001)

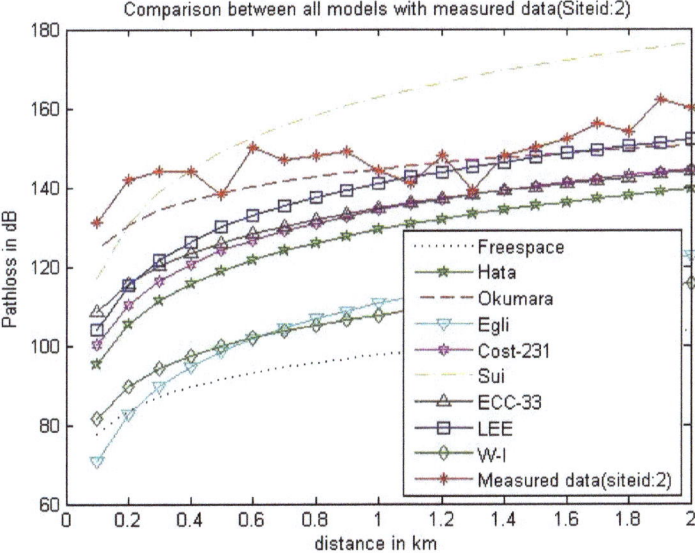

Figure 4.6 Comparison of Field Measured Path Loss and Predicted Path Loss with Distance (Site id NNL002)

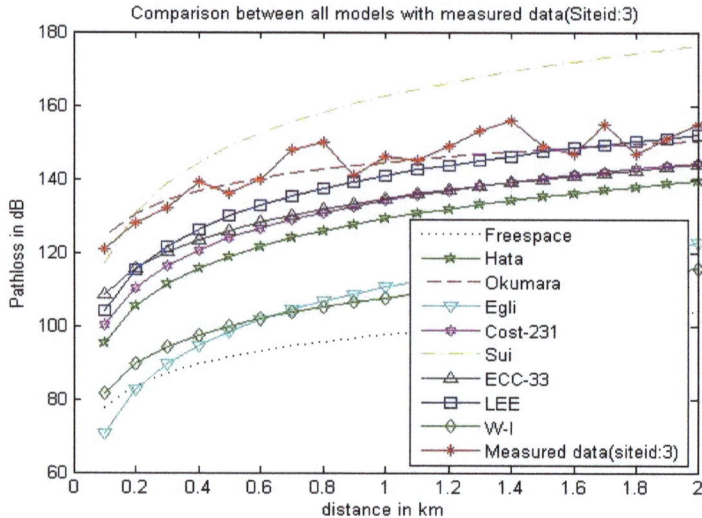

Figure 4.7 Comparison of Field Measured Path Loss and Predicted Path Loss with Distance (Site id NNL003)

The analysis of above figures 4.5, 4.6 and 4.7 shows that there is a noticeable difference between measured values and the predicted values of path loss. Each sample-point as shown in figure 4.5, 4.6 and 4.7 has been collected from all predefined routes of the three base stations. Around 20 to 30 sample-points of the received signal strength and T-R separation distance are recorded for each base station [111]. From figure 4.5, 4.6 and 4.7, it has been understood that propagation models does not predict precisely for all types of propagation environment. It is due to the fact that these models are designed in different terrains and the propagation of radio waves in built up areas is strongly influenced by the nature of the environment and in particular the size and density of buildings and the climatic conditions of that particular area [38].

The Free Space model and the field measured data are compared in the following section.

4.5 COMPARATIVE ANALYSIS BETWEEN FREE SPACE PATH LOSS MODEL AND FIELD MEASURED PATH LOSS

The formula for Free Space path loss is given as [154]:

$$PL\ (dB) = 32.45 + 20\ \log_{10}(f) + 20\ \log_{10}(d) \qquad (4.4)$$

Here, f is the operating frequency in Hz and d is the distance between transmitter and receiver in meters.

The equation for Free Space path loss model has been simulated with the help of MATLAB and these simulated results are compared with the practical field measured data which are recorded during drive test at three different BTS sites. The figure 4.8 shows the MATLAB file (M-file) of Free Space path loss model equation.

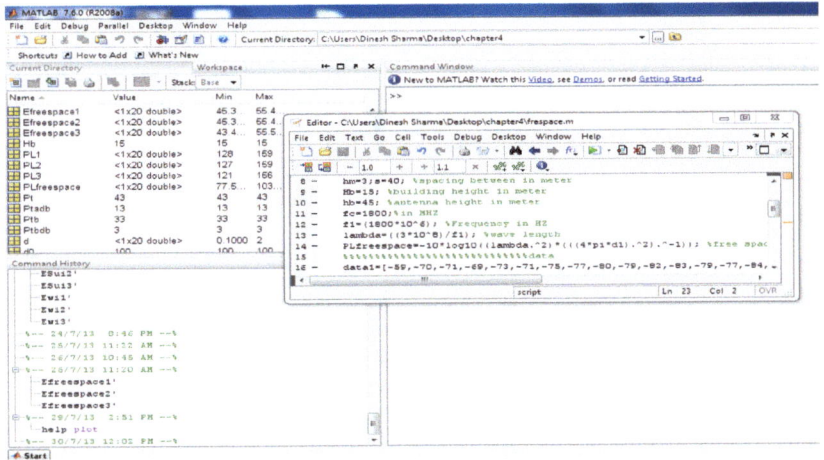

Figure 4.8 M-file of Free Space Path Loss Model

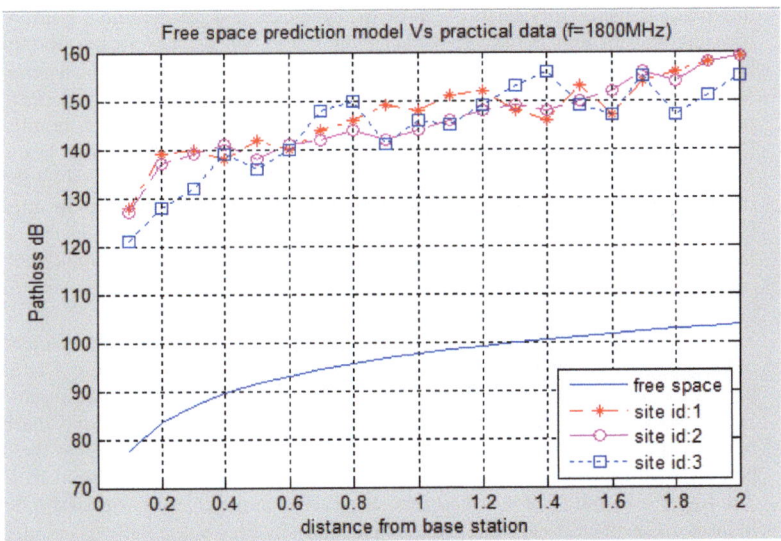

Figure 4.9 Comparison between Field Measured Data and Free Space Path Loss Model

93

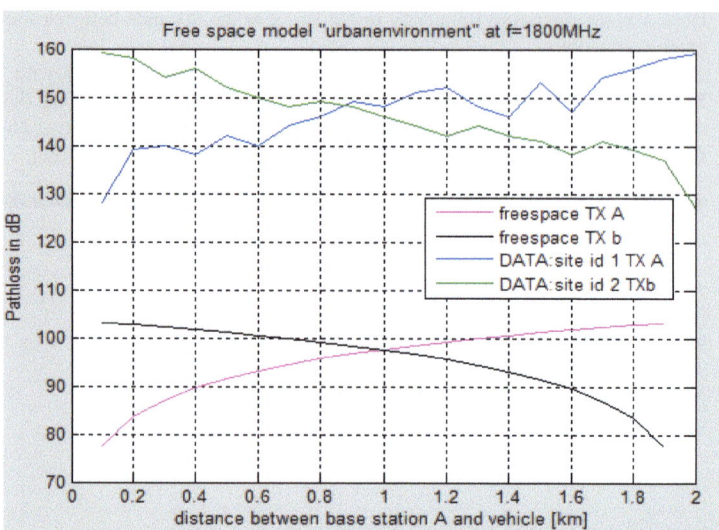

Figure 4.10 Variation of Path Loss between Free Space Path Loss and Practical Field Data for Two Adjacent Cells

The figure 4.9 depicts the variation of measured path loss with the Free Space path loss model. By analyzing the figure it is found that there is a significant difference around 40dB exists between the predicted and the measured path loss. The Free Space path loss is used to yield better results when there are no obstacles and hindrances on the path. It does not include factors such as the gain of the antennas used at the transmitter and receiver, nor any loss associated with hardware imperfections. In practical, the base station (BS) is located at a height of 45 meter and mobile station (MS) is often located in between the buildings or in street, so there is no line of sight path available. Therefore, the received signal at the MS end is influenced by the effect of reflection, diffraction and scattering. This entire phenomenon leads to the attenuation in the received signal. This explains the increased path loss observed in the currently taken field data measurement as compared to the Free Space path loss prediction.

The figure 4.10 shows the variation of path loss due the free space path loss model and the field measured data of two adjacent cells ids. The figure 4.10 helps in deciding the process of handoff in that particular area. After analyzing the figure 4.10, it is found that the handoff initiation point is 800m by measured data and the handoff point by the Free Space model is nearly 1 km. As highlighted earlier, that the handoff process is an expensive process to execute, so unnecessary handoffs should be avoided. The figure 4.11 and table 4.3 shows the error between the currently measured field data in three different cells ids and the predicted values by Free Space path loss model.

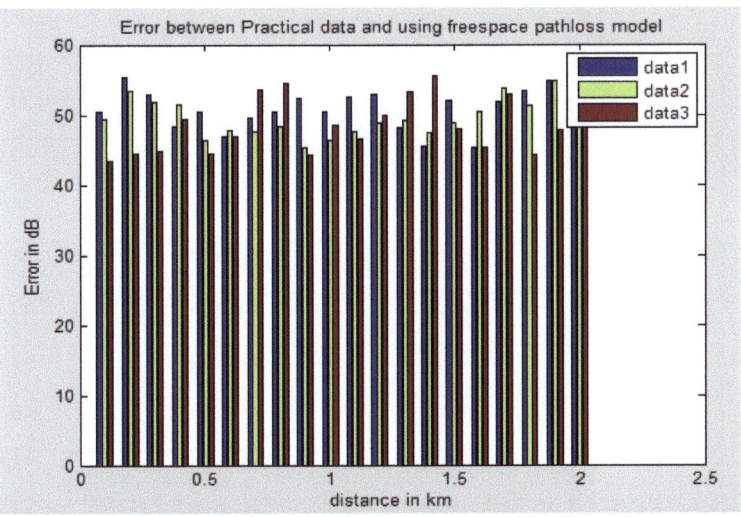

4.11 Variation of Error Between Field Measured Data and Free Space Path Loss Model

The W-I model and the field measured data are compared in the following section.

4.6 COMPARATIVE ANALYSIS BETWEEN W-I PATH LOSS MODEL AND FIELD MEASURED PATH LOSS

The equation of the Walfish-Ikegami (W-I) model is expressed [75] as:

In case of LOS, path loss is defined as:

$$PL_{LOS} = 42.6 + 26\log(d) + 20\log(f) \qquad (15)$$

And for NLOS condition

$$PL_{NLOS} = \{LFSL + Lrts + Lmsd\} \qquad (16)$$

This is the extended version of COST-231 and consist rooftop losses and building losses. The path loss equation of Walfish-Ikegami model has been simulated with the help of MATLAB as shown in figure 4.12 and the results obtained by this analysis are compared by currently measured field data.

Table 4.3 Error between measured and free space path loss model

Distance from base station (km)	Error (dB) Site id:NNL001	Error (dB) Site id:NNL002	Error (dB) Site id:NNL003
0.1	50.4528	49.4528	43.4528
0.2	55.4322	53.4322	44.4322
0.3	52.9104	51.9104	44.9104
0.4	48.4116	51.4116	49.4116
0.5	50.4734	46.4734	44.4734
0.6	46.8898	47.8898	46.8898
0.7	49.5508	47.5508	53.5508
0.8	50.3910	48.3910	54.3910
0.9	52.3679	45.3679	44.3679
1.0	50.4528	46.4528	48.4528
1.1	52.6249	47.6249	46.6249
1.2	52.8692	48.8692	49.8692
1.3	48.1739	49.1739	53.1739
1.4	45.5302	47.5302	55.5302
1.5	51.9310	48.9310	47.9310
1.6	45.3704	50.3704	45.3704
1.7	51.8438	53.8438	52.8438
1.8	53.3473	51.3473	44.3473
1.9	54.8777	54.8777	47.8777
2.0	55.4322	55.4322	51.4322

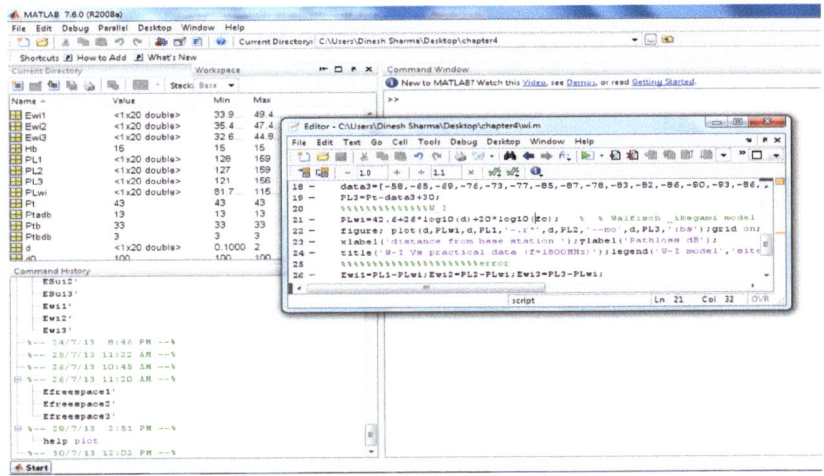

Figure 4.12 M-file of W-I Path Loss Model

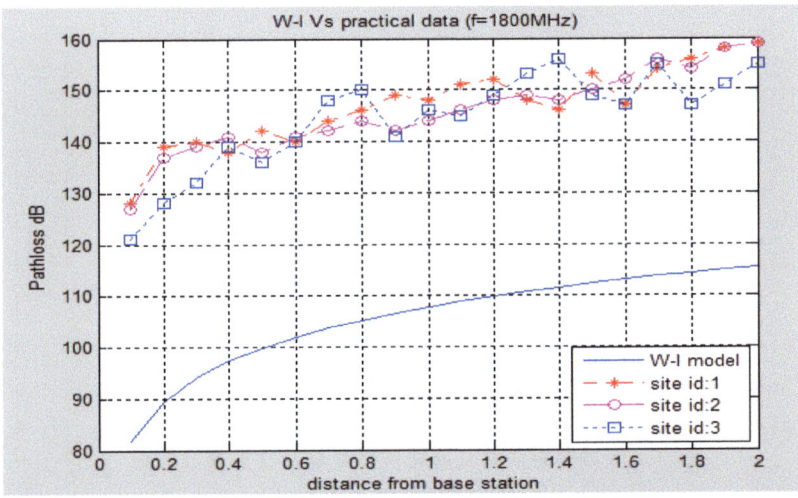

Figure 4.13 Comparison between Field Measured Path Loss and W-I Path Loss Model

After analyzing the figure 4.13 it has been observed that a difference of nearly 34 dB exists between the field measured data and the path loss values predicted by W-I path loss model. Theoretically the model gives good approximation to the path loss calculations. The W-I model is good for the suburban area, but not so affective for urban and rural area [75].

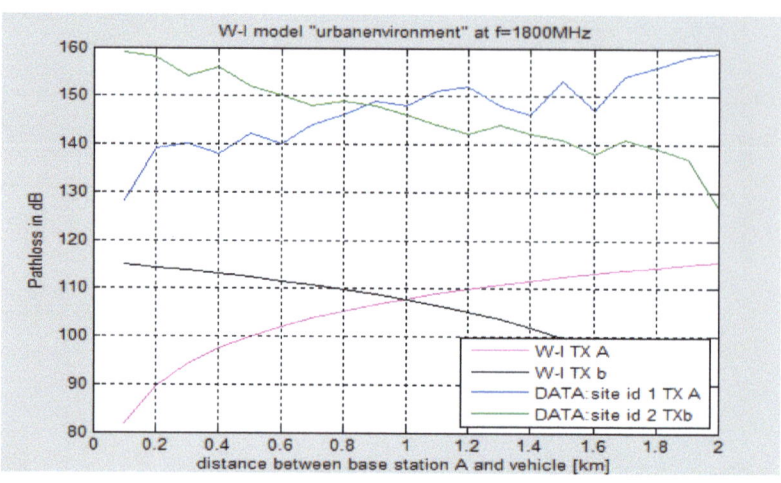

Figure 4.14 Variation of Path Loss between W-I Path Loss and Practical Field Data for Two Adjacent Cells

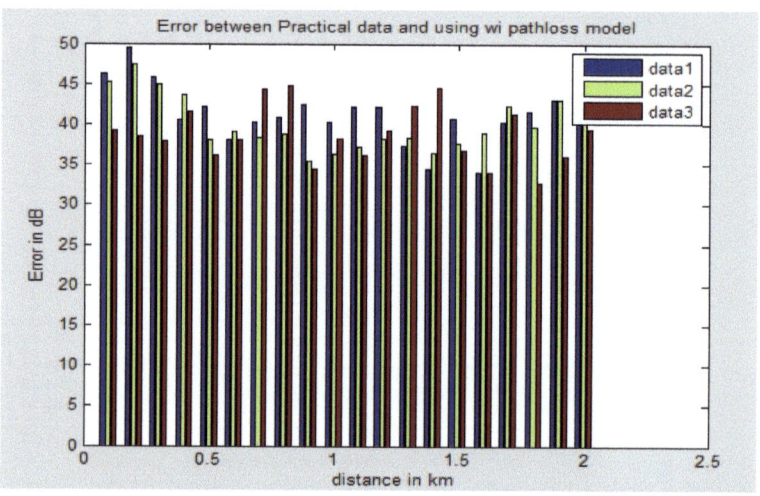

4.15 Variation of Error between Field Measured Data and W-I Path Loss Model

       The figure 4.14 shows the variation of path loss between the W-I path loss model and the currently recorded field data of two adjacent cells ids. The figure 4.14 helps in deciding the process of handoff in that particular area. After analyzing the figure 4.14, it is found that the handoff initiation point is 900m by currently measured data and the handoff point by the W-I model is nearly 1 km. The figure 4.15 and table 4.4 depicts the error between the field measured data in three different cells ids and the predicted values by W-I path loss model.

Table 4.4 Error between measured and W-I path loss model

Distance from base station (km)	Error(dB) Site id:NNL001	Error(dB) Site id:NNL002	Error(dB) Site id:NNL003
0.1	46.2945	45.2945	39.2945
0.2	49.4678	47.4678	38.4678
0.3	45.8894	44.8894	37.8894
0.4	40.6410	43.6410	41.6410
0.5	42.1213	38.1213	36.1213
0.6	38.0626	39.0626	38.0626
0.7	40.3220	38.3220	44.3220
0.8	40.8142	38.8142	44.8142
0.9	42.4842	35.4842	34.4842
1.0	40.2945	36.2945	38.2945
1.1	42.2183	37.2183	36.2183
1.2	42.2358	38.2358	39.2358
1.3	37.3320	38.3320	42.3320
1.4	34.4952	36.4952	44.4952
1.5	40.7162	37.7162	36.7162
1.6	33.9874	38.9874	33.9874
1.7	40.3029	42.3029	41.3029
1.8	41.6575	39.6575	32.6575
1.9	43.0470	43.0470	36.0470
2.0	43.4678	43.4678	39.4678

The LEE path loss model and the field measured data are compared in the following section.

4.7. COMPARATIVE ANALYSIS BETWEEN LEE PATH LOSS MODEL AND FIELD MEASURED PATH LOSS

The path loss equation of Lee model is given by:

$$L_P(dbm) = \begin{cases} 89 + 43.5\log_{10}\left(\dfrac{r}{1.6km}\right) + 10n\log_{10}\left(\dfrac{f}{900MHz}\right) - \alpha_0 & \text{Rural} \\ 101.7 + 38.4\log_{10}\left(\dfrac{r}{1.6km}\right) + 10n\log_{10}\left(\dfrac{f}{900MHz}\right) - \alpha_0 & \text{Suburban} \\ 110 + 36.8\log_{10}\left(\dfrac{r}{1.6km}\right) + 10n\log_{10}\left(\dfrac{f}{900MHz}\right) - \alpha_0 & \text{Urban} \end{cases}$$

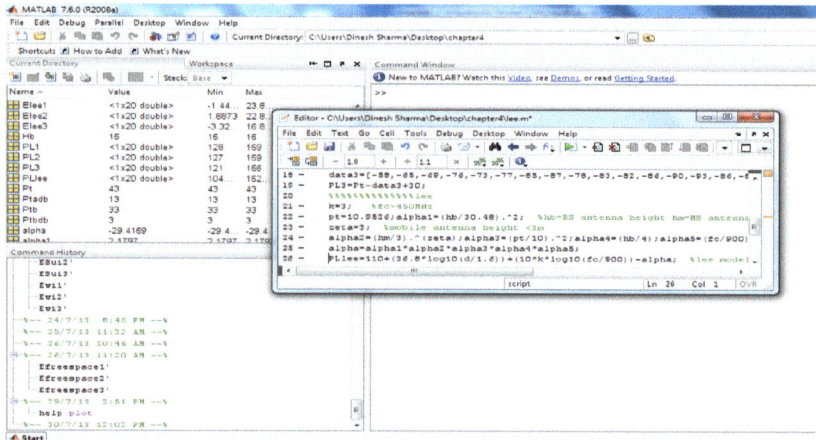

Figure 4.16 M-file of Lee Path Loss Model

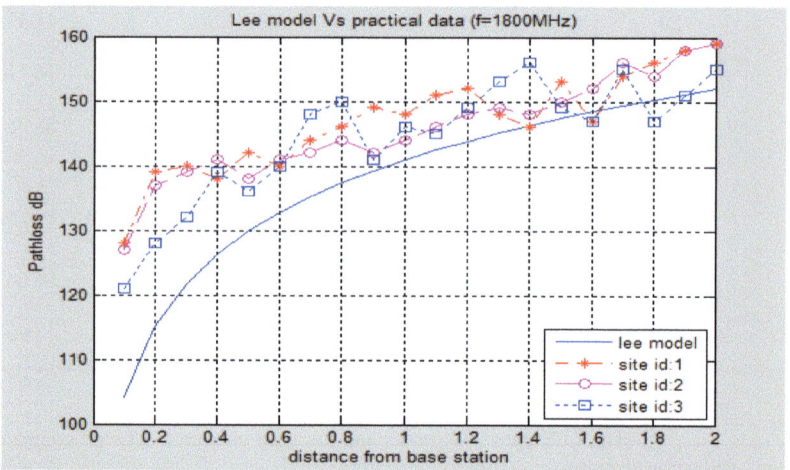

Figure 4.17 Comparison between Field Measured Path Loss and Lee Path Loss Model

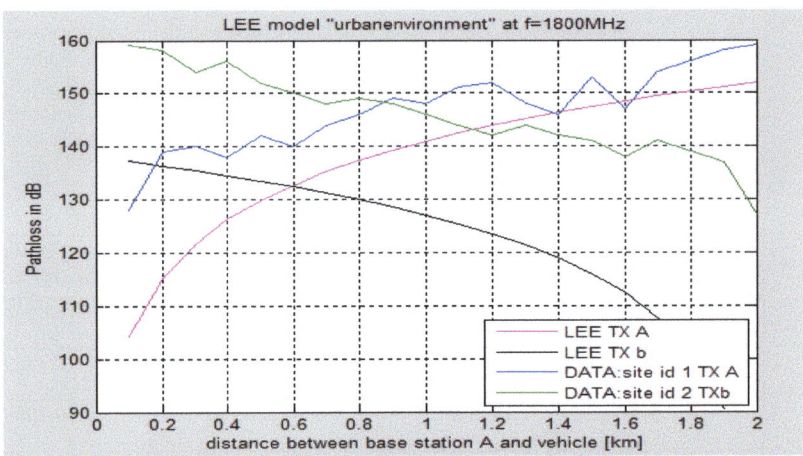

Figure 4.18 Variation of path loss between Lee path loss and Practical field data for two adjacent cells

      The figure 4.16 shows the M-file of Lee path loss model and with the help of this programming, a simulation has been done and the simulated results were compared with the field measured data. The comparison between the predicted value and the field measured data is shown in figure 4.17. The difference between the predicted value and the measured value is nearly 20 dB. The Lee model is applicable for only 150 MHz – 1500 MHz and this model is not good for hilly terrain [75]. So, for applying this model a correction must be needed to fill this significant gap.

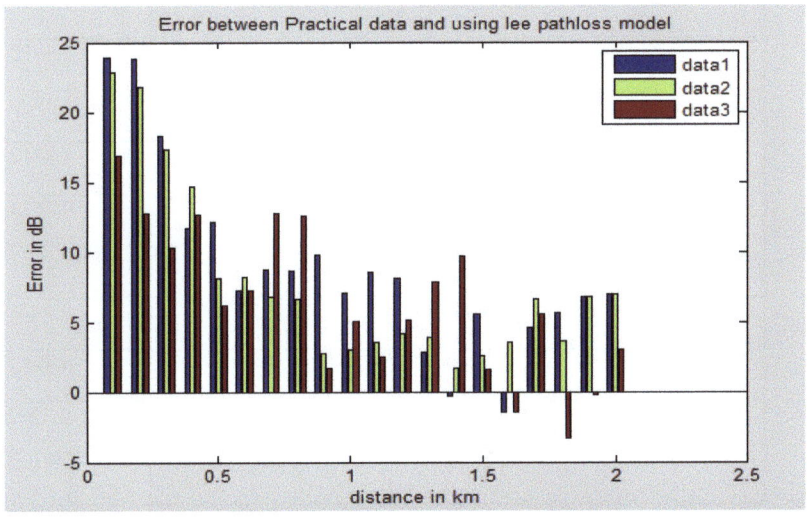

4.19 Variation of Error between Field Measured Data and Lee Path Loss Model

After observing figure 4.18, it is highlighted that the handoff process starts from the distance of 600m. But practically the handoff starts from 900m. So by keeping in the mind of handoff study, the Lee model is not advisable in this particular area. The figure 4.19 and the table 4.5 give you an idea about the error between the predicted value and the currently measured field data. The error varies between 1 dB to 17 dB at different points in the range of 2 km from the transmitter.

Table 4.5 Error between measured and Lee path loss model

Distance from base station (km)	Error (dB) Site id:NNL001	Error (dB) Site id:NNL002	Error (dB) Site id:NNL003
0.1	23.8648	22.8648	16.8648
0.2	23.7869	21.7869	12.7869
0.3	18.3068	17.3068	10.3068
0.4	11.7090	14.7090	12.7090
0.5	12.1427	8.1427	6.1427
0.6	7.2289	8.2289	7.2289
0.7	8.7652	6.7652	12.7652
0.8	8.6311	6.6311	12.6311
0.9	9.7487	2.7487	1.7487
1.0	7.0648	3.0648	5.0648
1.1	8.5416	3.5416	2.5416
1.2	8.1509	4.1509	5.1509
1.3	2.8717	3.8717	7.8717
1.4	-0.3127	1.6873	9.6873
1.5	5.5847	2.5847	1.5847
1.6	-1.4468	3.5532	-1.4468
1.7	4.5843	6.5843	5.5843
1.8	5.6708	3.6708	-3.3292
1.9	6.8067	6.8067	-0.1933
2.0	6.9869	6.9869	2.9869

The Egli path loss model and the field measured data are compared in the following section.

4.8 COMPARATIVE ANALYSIS BETWEEN EGLI PATH LOSS MODEL AND FIELD MEASURED PATH LOSS

The formula for Egli path loss model is expressed as [186]:

$$PL = G_B G_M \left(\frac{h_B h_M}{d^2}\right)^2 \left(\frac{40}{f}\right)^2 \quad (4.6)$$

or

$$PL_{egli} = 117 + 40\log_{10}D + 20\log_{10}fc - 20\log_{10}(hb1 \times hm1)$$

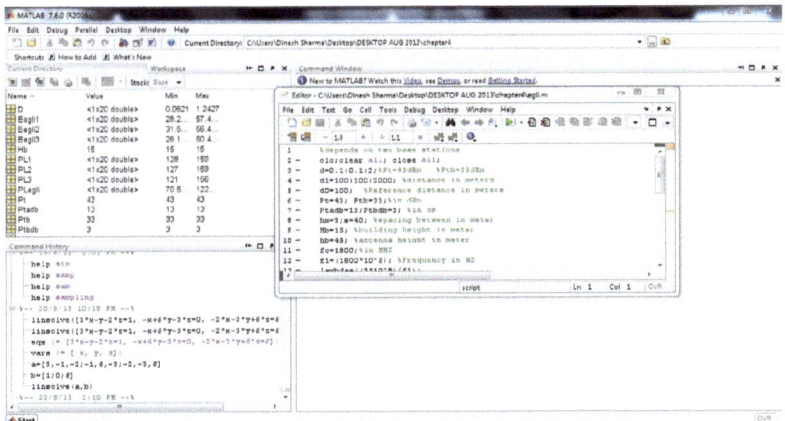

Figure 4.20 M-file of Egli Path Loss Model

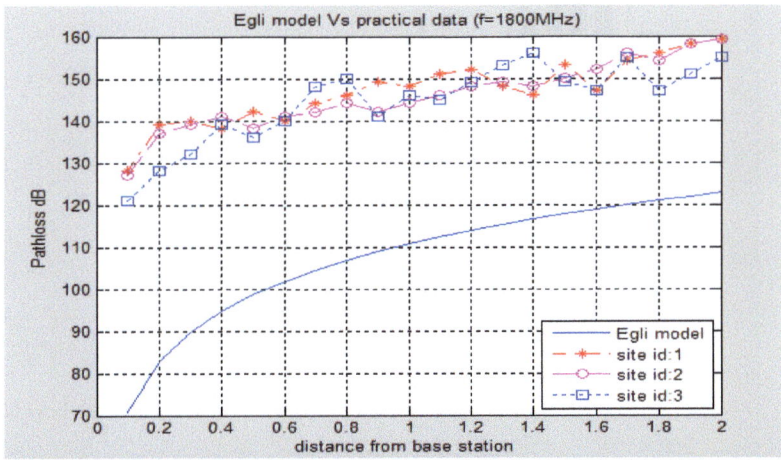

Figure 4.21 Comparison between Field Measured Path Loss and Egli Path Loss Model

The equation for Egli path loss model has been simulated with the help of MATLAB and these simulated results are compared with the practical field measured data which are presently recorded during drive test at three different BTS sites. The figure 4.20 shows the M-file of free space path loss model equation. The Egli model is good for the 900 MHz frequency band. The comparative analysis of the currently measured field data and the calculated values by the Egli model is shown in figure 4.21.

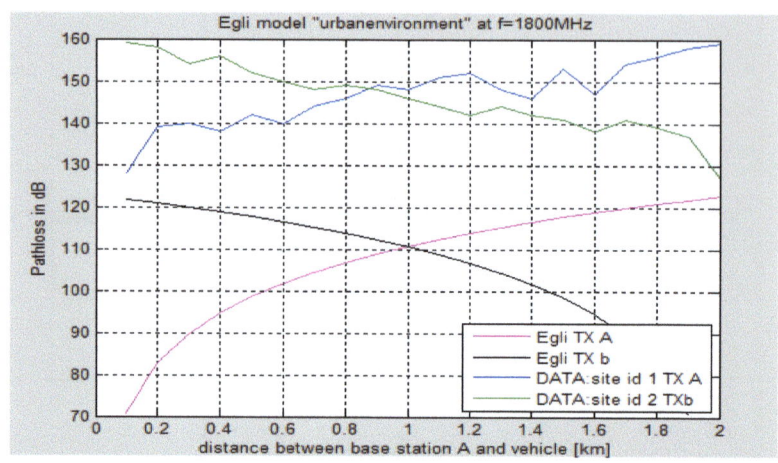

Figure 4.22 Variation of Path Loss between Egli Path Loss and Practical Field Data for Two Adjacent Cells

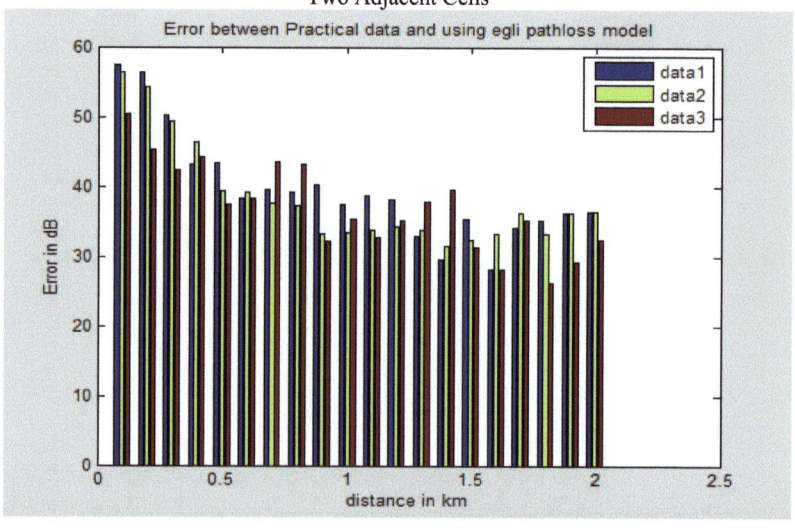

4.23 Variation of Error between Field Measured Data and Egli Path Loss Model

Analysis of figure 4.21 and 4.23 describes that measured result has more attenuation around the near field (Error is nearly 50 dB). This may be due to the shadowing effect of antenna height. As the distance between BS and MS increases, difference between measured path loss and Egli prediction reduces, since Egli path loss increases faster with the T-R separation. This can be explained by the nature of Egli model. Basically Egli model is a terrain model and this type of model provide a measure of path loss as a function of only distance and terrain roughness. Egli model does not include the effect of clutter in the prediction of path loss. Due to this it does not provide a precise result for mobile link but it is useful for planning of fixed link. The figure 4.22 describes the distance at which the handoff process starts.

Now, the comparison of Bertoni path loss model and the field measured data is given in the following section.

4.9 COMPARATIVE ANALYSIS BETWEEN BERTONI PATH LOSS MODEL AND FIELD MEASURED PATH LOSS [228]

The path loss equation for Bertoni model is given as:
$$L = P_0 Q^2 P_1, \tag{4.7}$$

Here P_0 is the Free Space Path loss for Omni directional antennas, Q^2 reflects the signal power reduction due to buildings that block the receiver at street level, and P_1 is based on the signal loss from the rooftop to the street due to diffraction. The model has been adopted for the IMT-2000 standard.

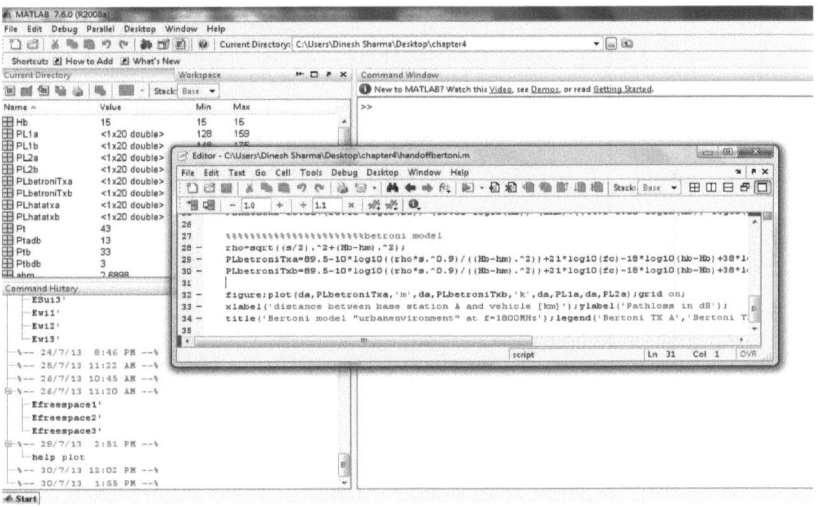

Figure 4.24 M-file of Bertoni Path Loss Model

Table 4.6 Error between measured and Egli path loss model

Distance from base station (km)	Error(dB) Site id:NNL001	Error(dB) Site id:NNL002	Error(dB) Site id:NNL003
0.1	57.4066	56.4066	50.4066
0.2	56.3654	54.3654	45.3654
0.3	50.3217	49.3217	42.3217
0.4	43.3242	46.3242	44.3242
0.5	43.4478	39.4478	37.4478
0.6	38.2805	39.2805	38.2805
0.7	39.6027	37.6027	43.6027
0.8	39.2830	37.2830	43.2830
0.9	40.2369	33.2369	32.2369
1.0	37.4066	33.4066	35.4066
1.1	38.7509	33.7509	32.7509
1.2	38.2393	34.2393	35.2393
1.3	32.8488	33.8488	37.8488
1.4	29.5615	31.5615	39.5615
1.5	35.3629	32.3629	31.3629
1.6	28.2418	33.2418	28.2418
1.7	34.1886	36.1886	35.1886
1.8	35.1957	33.1957	26.1957
1.9	36.2564	36.2564	29.2564
2.0	36.3654	36.3654	32.3654

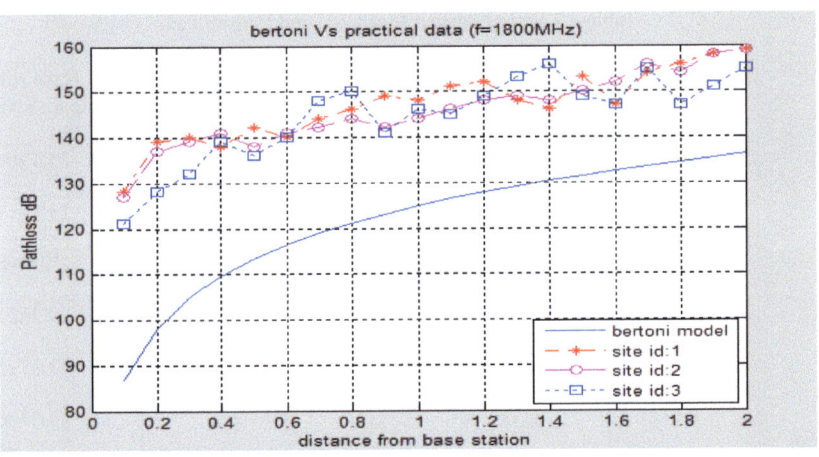

Figure 4.25 Comparison between Field Measured Path Loss and Bertoni Path Loss Model

The path loss equation of Bertoni model has been simulated with the help of MATLAB and the M-file of path loss model is shown in figure 4.24. The comparative analysis between the field measured data and the predicted values is shown in figure 4.25. After analyzing the comparative analysis it has been observed that initially the difference between the predicted and the measured value is 34 dB and this difference decreases up to 20 dB and it is shown in figure 4.27 and table 4.7. The figure 4.26 depicts the variation of path loss between predicted values and the currently measured field data of two adjacent cells. This analysis helps in deciding the initiation of handoff process.

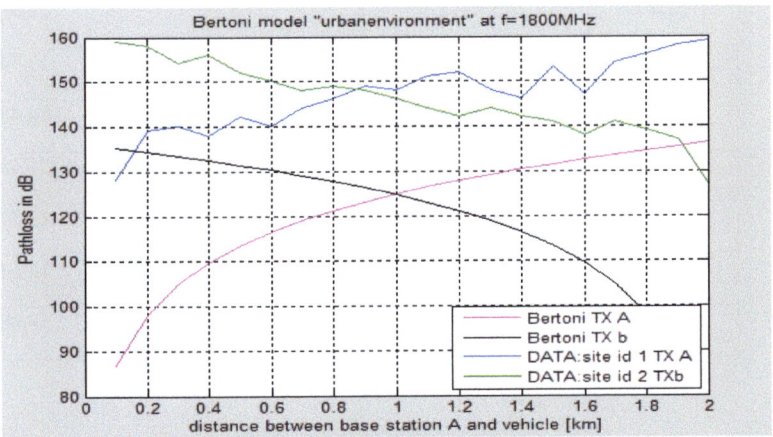

4.26 Variation of Path Loss between Bertoni Path Loss and Practical Field Data for Two Adjacent Cells

Table 4.7 Error between Measured and Bertoni Path Loss Model

Distance from base station (km)	Error (dB) Site id:NNL001	Error (dB) Site id:NNL002	Error (dB) Site id:NNL003
0.1	41.2404	40.2404	34.2404
0.2	40.8012	38.8012	29.8012
0.3	35.1098	34.1098	27.1098
0.4	28.3621	31.3621	29.3621
0.5	28.6795	24.6795	22.6795
0.6	23.6706	24.6706	23.6706
0.7	25.1266	23.1266	29.1266
0.8	24.9229	22.9229	28.9229
0.9	25.9792	18.9792	17.9792
1.0	23.2404	19.2404	21.2404
1.1	24.6674	19.6674	18.6674
1.2	24.2315	20.2315	21.2315
1.3	18.9105	19.9105	23.9105
1.4	15.6875	17.6875	25.6875
1.5	21.5489	18.5489	17.5489
1.6	14.4838	19.4838	14.4838
1.7	20.4833	22.4833	21.4833
1.8	21.5400	19.5400	12.5400
1.9	22.6477	22.6477	15.6477
2.0	22.8012	22.8012	18.8012

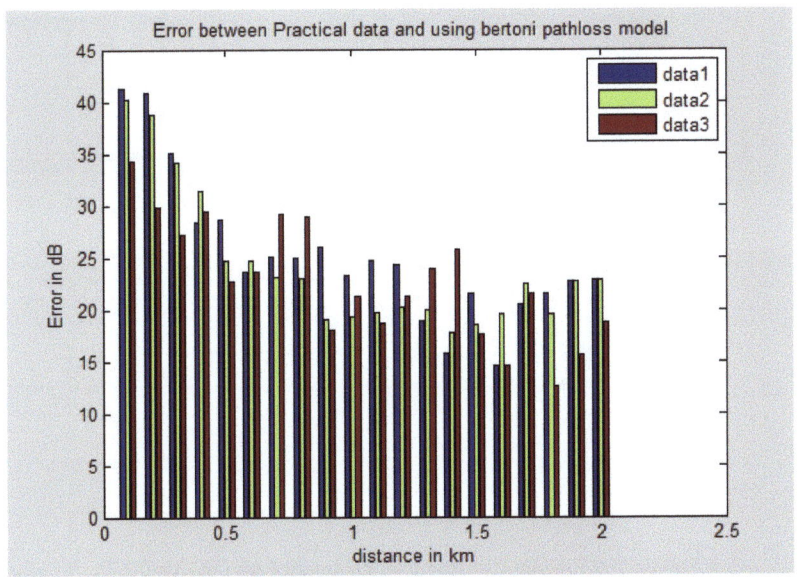

4.27 Variation of Error between Field Measured Data and Bertoni Model

The Okumura path loss model and the currently measured field data are compared in the following section.

4.10 COMPARATIVE ANALYSIS BETWEEN OKUMURA PATH LOSS MODEL AND FIELD MEASURED PATH LOSS

The Okumura model is expressed as [141,157]:

$$PL\ (dB) = L_F + A_{mu}\ (f,d) - G\ (h_{te}) - G\ (h_{re}) - G_{area} \qquad (4.7)$$

Where, PL(dB) is the propagation path loss, L_F is free space path loss, A_{mu} is median attenuation relative to free space attenuation, $G(h_{te})$ is the base station antenna height gain factor, $G(h_{re})$ is mobile antenna height gain factor, and G_{area} is gain corresponding to specific environment.

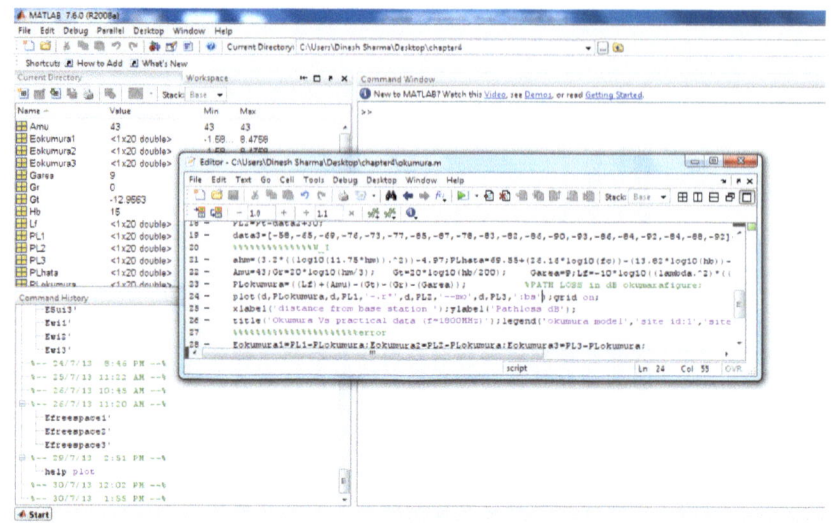

Figure 4.28 M-file of Okumura Path Loss Model

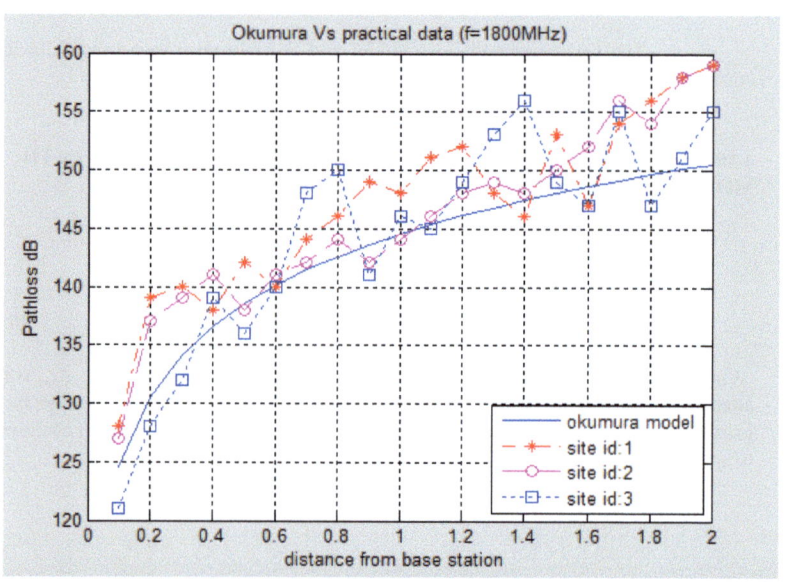

Figure 4.29 Comparison between Field Measured Path Loss and Okumura Path Loss Model

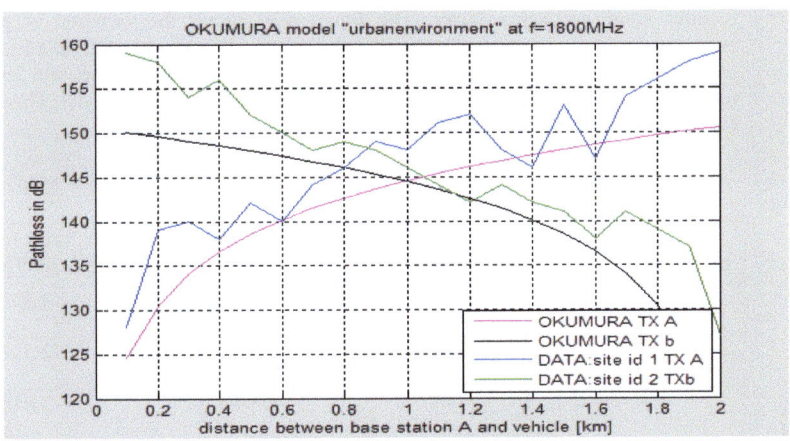

Figure 4.30 Variation of Path Loss between Okumura Path Loss Model and Practical Field Data for Two Adjacent Cells

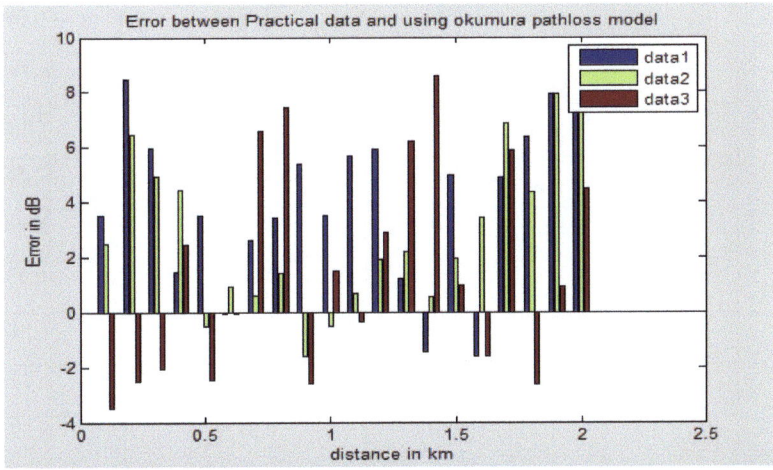

Figure 4.31 Variation of Error between Field Measured Data and Okumura Path Loss Model

The figure 4.28 illustrates the M-file of Okumura model, used to simulate the path loss equation. Figure 4.29 shows the comparative analysis of currently measured field data and the predicted path loss values by Okumura model. As the difference (between the predicted values and the measured values) is very small and in acceptable range, the model indicates a good agreement with the measured path loss. This is because Okumura model takes all the parameter into account such as free space loss, median attenuation relative to free space, transmitting/receiving antenna gain and gain due to environment, but this model is also does not give correct results. It means it requires some correction factor. The figure 4.30 shows the variation of path loss between the two adjacent cells which helps in deciding the handoff initiation. After analyzing this figure it is concluded that the distance of handoff initiation is same as measured by drive test and predicted values by Okumura path loss model. Therefore the Okumura path loss model gives better results in Narnaul (Haryana).

Table 4.8 Error between Measured Data and Okumura Path Loss Model

Distance from base station (km)	Error (dB) Site id:NNL001	Error (dB) Site id:NNL002	Error (dB) Site id:NNL003
0.1	3.4964	2.4964	-3.5036
0.2	8.4758	6.4758	-2.5242
0.3	5.9540	4.9540	-2.0460
0.4	1.4552	4.4552	2.4552
0.5	3.5170	-0.4830	-2.4830
0.6	-0.0666	0.9334	-0.0666
0.7	2.5945	0.5945	6.5945
0.8	3.4346	1.4346	7.4346
0.9	5.4116	-1.5884	-2.5884
1.0	3.4964	-0.5036	1.4964
1.1	5.6686	0.6686	-0.3314
1.2	5.9128	1.9128	2.9128
1.3	1.2176	2.2176	6.2176
1.4	-1.4261	0.5739	8.5739
1.5	4.9746	1.9746	0.9746
1.6	-1.5860	3.4140	-1.5860
1.7	4.8874	6.8874	5.8874
1.8	6.3910	4.3910	-2.6090
1.9	7.9214	7.9214	0.9214
2.0	8.4758	8.4758	4.4758

The Cost-231 path loss model and the field measured data are compared in the following section.

4.11 COMPARATIVE ANALYSIS BETWEEN COST 231 PATH LOSS MODEL AND FIELD MEASURED PATH LOSS

The European Cooperative for Scientific and Technical (COST) research extended the Hata model to 2 GHz as follows [186]:

$$PL\ (dB) = 46.3 + 33.9\ \log_{10}(f) - 13.82\ \log_{10}(h_{te}) - a(h_m) \\ + (44.9 - 6.55\ \log_{10}(h_{te}))\ \log_{10}d + c_m \quad (4.8)$$

Here $a(h_m)$ is the mobile antenna correction factor

for a small to medium sized city, it is defined as:

$$a(h_m) = (1.1\ \log f - 0.7)\ h_{re} - (1.56\ \log f - 0.8)\ dB \quad (4.9)$$

and for a large city, it is defined as:

$$a(h_m) = 8.29\ (\log 1.54 h_{re})^2 - 1.1\ dB \quad \text{for } f \leq 300\ \text{MHz} \quad (4.10)$$

$$a(h_m) = 3.2\ (\log 11.75 h_{re})^2 - 4.97\ dB \quad \text{for } f \geq 300\ \text{MHz} \quad (4.11)$$

and c_m is 0 dB for medium sized cities and suburban areas and it is 3 dB for urban areas.

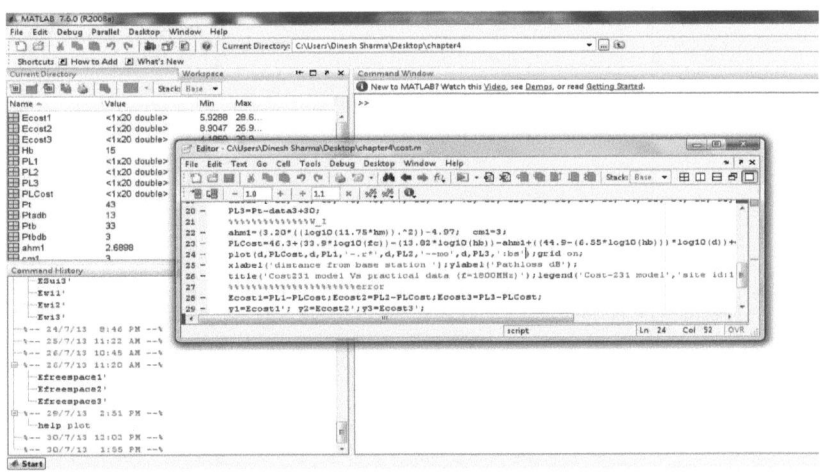

Figure 4.32 M-file of COST 231 Path Loss Model

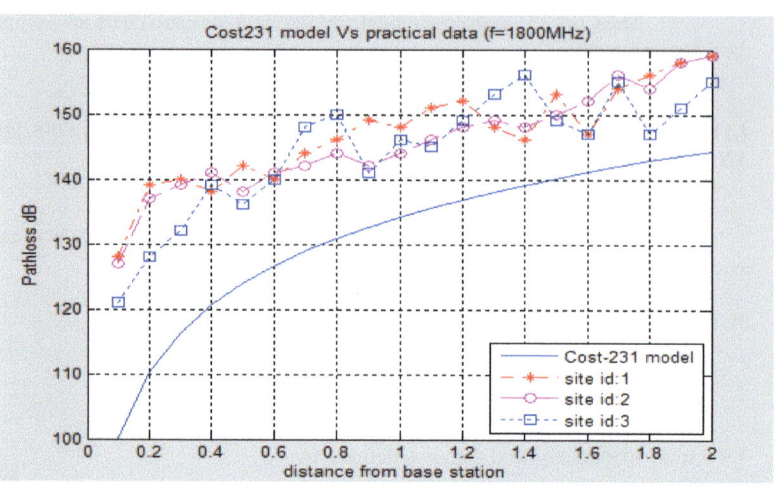

Figure 4.33 Comparison between Field Measured Path loss and COST 231 Path Loss Model

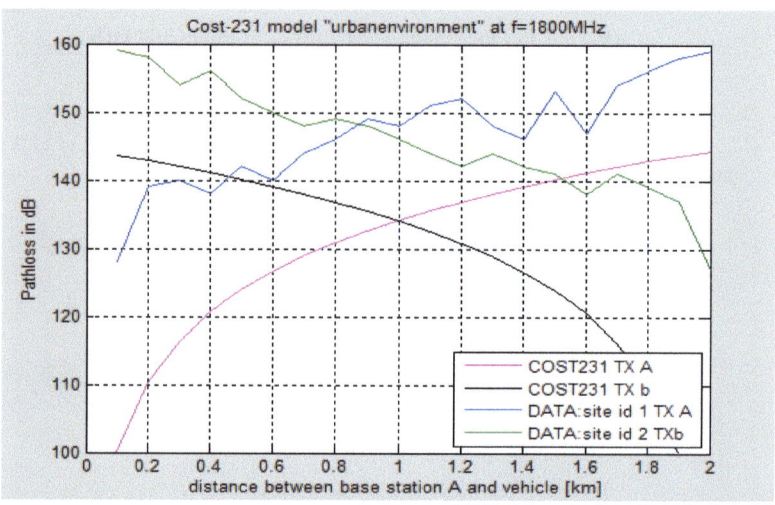

Figure 4.34 Variation of Path Loss between Cost 231 Model and Practical Field Data for Two Adjacent Cells

The equation for COST-231 path loss model has been simulated with the help of MATLAB and these simulated results were compared with the practical field measured data which were presently collected during drive test at three different BTS sites. The figure 4.33 shows the M-file of Cost 231 path loss model equation.

For the analysis of Cost 231 model, consider the figure 4.33, 4.35 and table 4.9. The measured path loss and the COST 231 path loss model prediction for the region shows that the deviation of the model from the measured path loss is nearly 20dB. This difference is due to considering the impact of diffraction from rooftops, building and scatter loss. At 1800 meter the deviation reached to a maximum value of around 4dB.

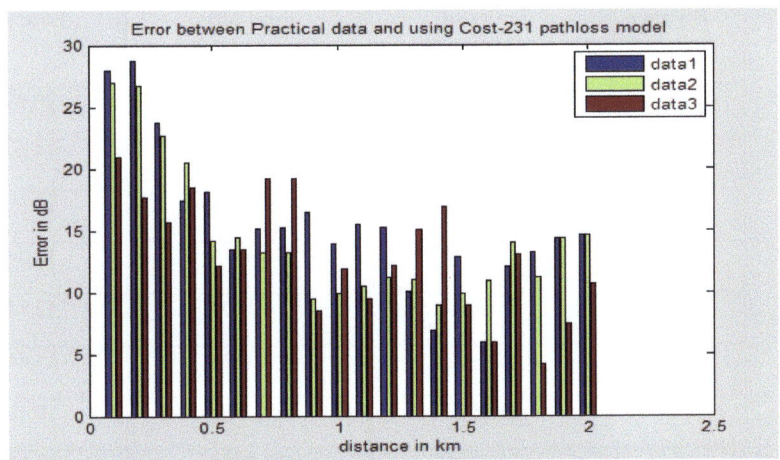

Figure 4.35 Variation of Error between Field Measured Data and Cost 231 Model

The ECC 33 path loss model and the field measured data are compared in the following section.

4.12 COMPARATIVE ANALYSIS BETWEEN ECC33 PATH LOSS MODEL AND FIELD MEASURED PATH LOSS

The ECC 33 path loss is defined as [186] [145]:

$$PL(dB) = A_{fs} + A_{bm} - G_b - G_r \qquad (4.12)$$

Here,
A_{fs} is Free space attenuation
$$= 92.4 + 20 \log_{10}(d) + 20 \log_{10}(f) \qquad (4.13)$$

A_{bm} is basic median path loss
$$= 20.41 + 9.83 \log_{10}(d) + 7.894 \log_{10}(f) + 9.56[\log_{10}(f)]^2 \qquad (4.14)$$

G_b is BS height gain factor
$$= \log_{10}(h_b/200)\{13.958 + 5.8[\log_{10}(d)]^2\} \qquad (4.15)$$

G_r is terminal height gain factor

$$= [42.57 + 13.7 \log_{10}(f)][\log_{10}(h_r) - 0.585] \qquad (4.16)$$

Table 4.9 Error between Measured and Cost 231 Path Loss Model

Distance from base station (km)	Error (dB) Site id:NNL001	Error (dB) Site id:NNL002	Error (dB) Site id:NNL003
0.1	27.9550	26.9550	20.9550
0.2	28.6984	26.6984	17.6984
0.3	23.6987	22.6987	15.6987
0.4	17.4419	20.4419	18.4419
0.5	18.1400	14.1400	12.1400
0.6	13.4422	14.4422	13.4422
0.7	15.1612	13.1612	19.1612
0.8	15.1854	13.1854	19.1854
0.9	16.4425	9.4425	8.4425
1.0	13.8835	9.8835	11.8835
1.1	15.4732	10.4732	9.4732
1.2	15.1857	11.1857	12.1857
1.3	10.0013	11.0013	15.0013
1.4	6.9047	8.9047	16.9047
1.5	12.8838	9.8838	8.8838
1.6	5.9288	10.9288	5.9288
1.7	12.0318	14.0318	13.0318
1.8	13.1860	11.1860	4.1860
1.9	14.3860	14.3860	7.3860
2.0	14.6270	14.6270	10.6270

The path loss equation of ECC-33 model has been simulated with the help of MATLAB and the M-File is shown in figure 4.36.

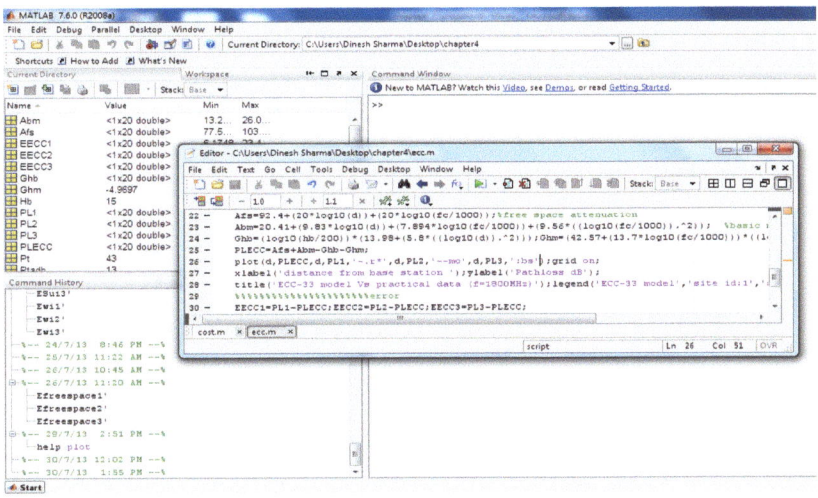

Figure 4.36 M-file of ECC 33 Path Loss Model

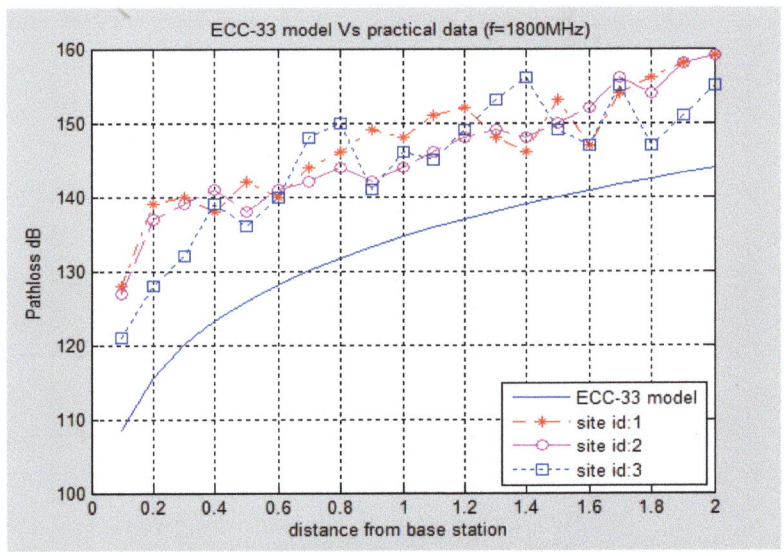

Figure 4.37 Comparison between Field Measured Path Loss and ECC-33 Path Loss Model

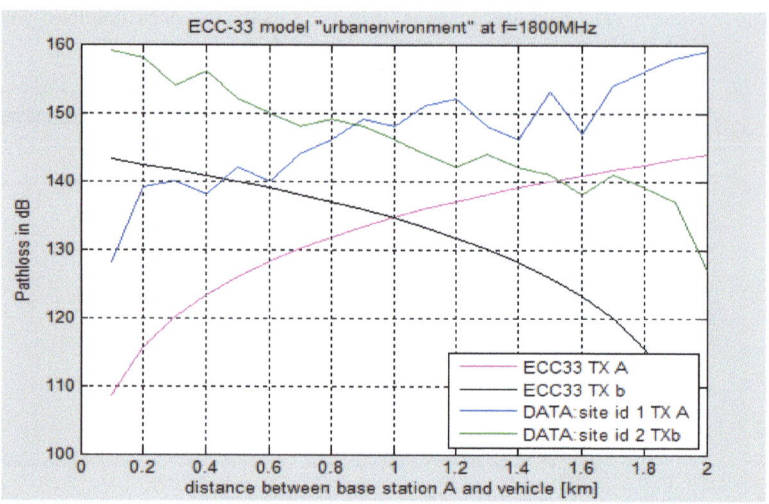

Figure 4.38 Variation of Path Loss between ECC-33 Path Loss Model and Practical Field Data for Two Adjacent Cells

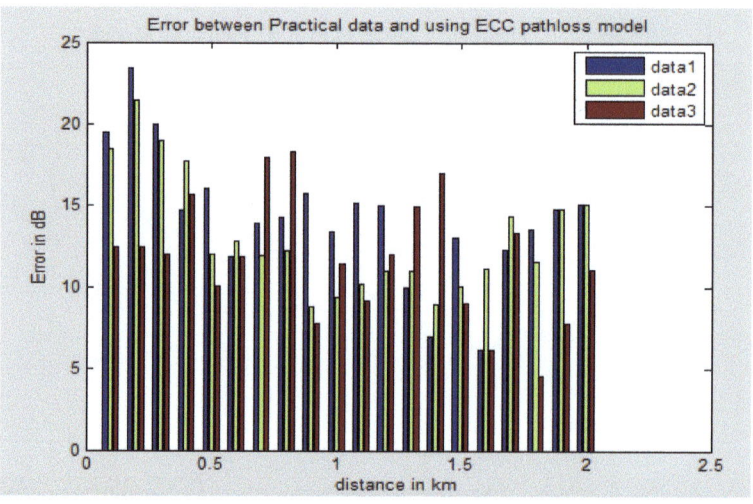

Figure 4.39 Variation of Error between Field Measured Data and ECC-33 Path Loss Model

The ECC-33 undershoots the measured value from the field. From figure 4.37, 4.39 and table 4.10, it is observed that the ECC-33 model does not give the accurate results. The variation of the error in the ECC-33 model and the measured field data varies from 4 to 18 dB. The figure 4.38 shows the variation of path loss of two adjacent cells ids.

Table 4.10 Error between measured and ECC-33 path loss model

Distance from base station (km)	Error(dB) Site id:NNL001	Error(dB) Site id:NNL002	Error(dB) Site id:NNL003
0.1	19.4930	18.4930	12.4930
0.2	23.4349	21.4349	12.4349
0.3	19.9905	18.9905	11.9905
0.4	14.6958	17.6958	15.6958
0.5	16.0595	12.0595	10.0595
0.6	11.8531	12.8531	11.8531
0.7	13.9509	11.9509	17.9509
0.8	14.2758	12.2758	18.2758
0.9	15.7774	8.7774	7.7774
1.0	13.4203	9.4203	11.4203
1.1	15.1791	10.1791	9.1791
1.2	15.0348	11.0348	12.0348
1.3	9.9726	10.9726	14.9726
1.4	6.9811	8.9811	16.9811
1.5	13.0510	10.0510	9.0510
1.6	6.1748	11.1748	6.1748
1.7	12.3465	14.3465	13.3465
1.8	13.5607	11.5607	4.5607
1.9	14.8131	14.8131	7.8131
2.0	15.1001	15.1001	11.1001

The SUI path loss model and the field measured data are compared in the following section.

4.13 COMPARATIVE ANALYSIS BETWEEN SUI PATH LOSS MODEL AND FIELD MEASURED PATH LOSS

The basic SUI path loss model equation with correction factors is represented as [211]:

$$PL = A + 10\gamma \log_{10}\left[\frac{d}{d_0}\right] + X_f + X_h + s \quad \textbf{for} \ d > d_0 \quad (4.17)$$

Here, d is the distance between the Access Points (AP) and the Customer Premises Equipment (CPE) antennas in meters, $d_0 = 100$ m and s is log normally distributed factor (used to account for the shadow fading owing to trees and other clutter) and has a value between 8.2 dB and 10.6 dB [76]. The figure 4.40 shows the M-file of SUI path loss model.

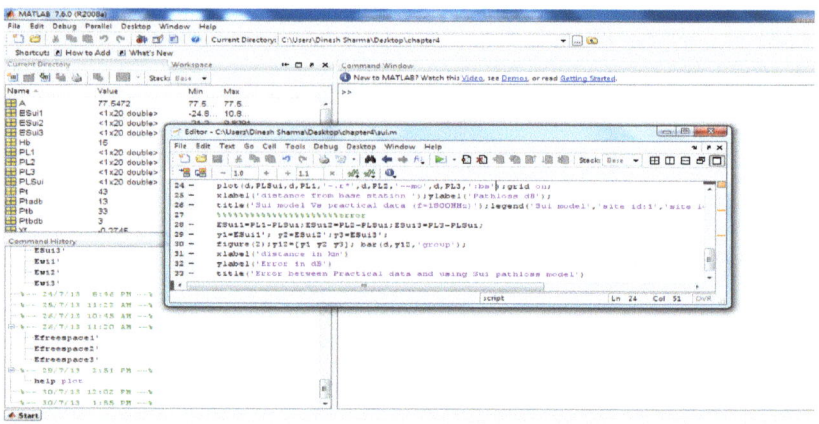

Figure 4.40 M-file of SUI Path Loss Model

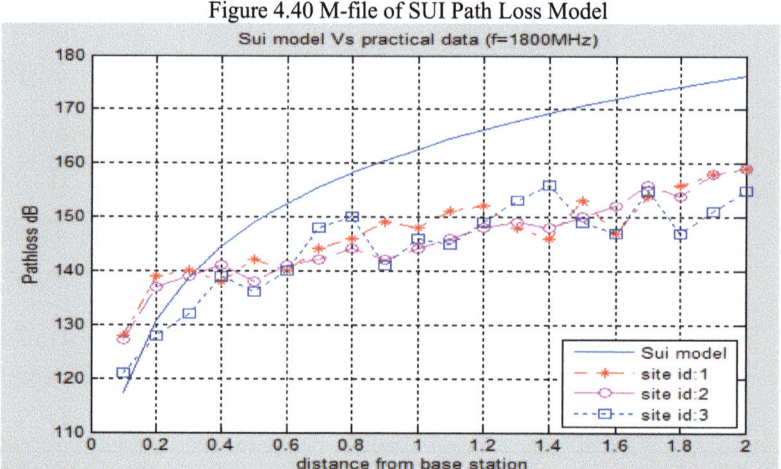

Figure 4.41 Comparison between Field Measured Path Loss and SUI Path Loss Model

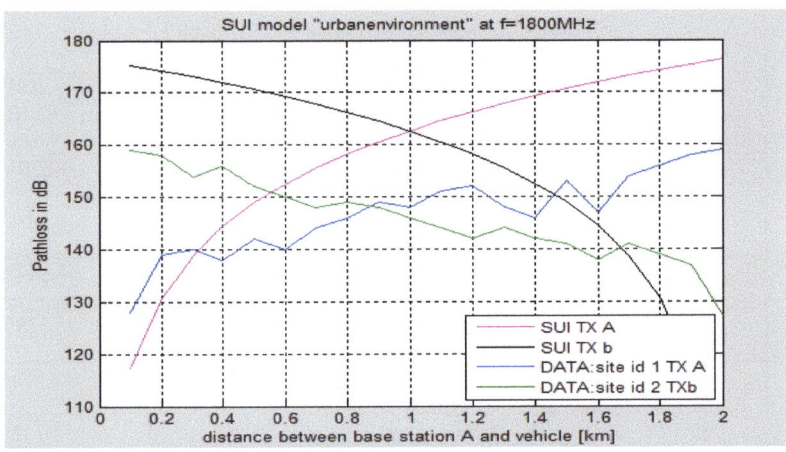

Figure 4.42 Variation of Path Loss between SUI Path Loss Model and Practical Field Data for Two Adjacent Cells

The result of the performance of the SUI model with the help of figure 4.41, 4.43 and table 4.11 is that this model overshoots the measured path loss. This model is designed for 2.5 GHz to 3.5 GHz for highly terrain with Moderate to heavy foliage densities. The error is nearly 28 dB in the predicted value and the measured value. The figure 4.42 shows the variation of the path loss of two adjacent cells. After analyzing this figure it has been observed that this model does not predict the path loss appropriately in Narnaul (Haryana).

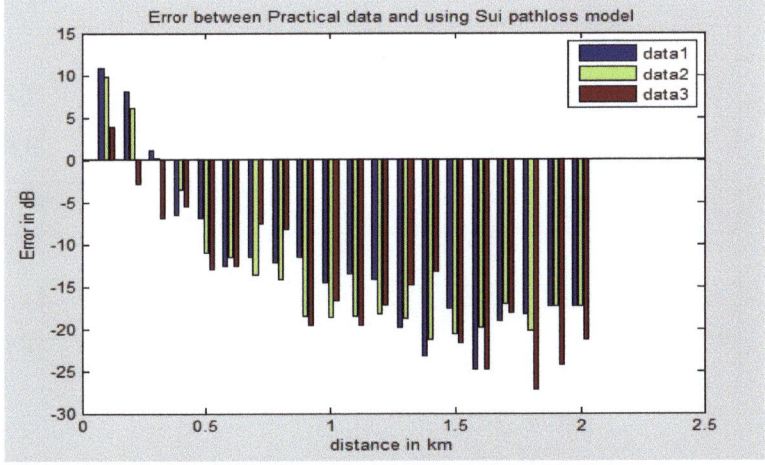

Figure 4.43 Variation of Error between Field Measured Data and SUI Path Loss Model

Table 4.11 Error between Measured and SUI Path Loss Model

Distance from base station (km)	Error(dB) Site id:NNL001	Error(dB) Site id:NNL002	Error(dB) Site id:NNL003
0.1	10.8291	9.8291	3.8291
0.2	8.1548	6.1548	-2.8452
0.3	1.1559	0.1559	-6.8441
0.4	-6.5195	-3.5195	-5.5195
0.5	-6.9216	-10.9216	-12.9216
0.6	-12.5184	-11.5184	-12.5184
0.7	-11.5595	-13.5595	-7.5595
0.8	-12.1938	-14.1938	-8.1938
0.9	-11.5174	-18.5174	-19.5174
1.0	-14.5959	-18.5959	-16.5959
1.1	-13.4762	-18.4762	-19.4762
1.2	-14.1927	-18.1927	-17.1927
1.3	-19.7718	-18.7718	-14.7718
1.4	-23.2338	-21.2338	-13.2338
1.5	-17.5948	-20.5948	-21.5948
1.6	-24.8680	-19.8680	-24.8680
1.7	-19.0640	-17.0640	-18.0640
1.8	-18.1916	-20.1916	-27.1916
1.9	-17.2583	-17.2583	-24.2583
2.0	-17.2702	-17.2702	-21.2702

The HATA path loss model and the field measured data are compared in the following section.

4.14. COMPARATIVE ANALYSIS BETWEEN HATA PATH LOSS MODEL AND FIELD MEASURED PATH LOSS

Hata model for path loss prediction in urban area is given below [146]:

$$PL_{pu(OH)} (dB) = 69.55 + 26.16 \log f_c - 13.82 \log h_{te} - a(h_m) + (44.9 - 6.55 \log h_{te})\log d \quad (4.18)$$

Where,
$PL_{pu(OH)}$ is path loss prediction for urban area in dB
f_c is carrier frequency in MHz
d is distance between base station and mobile station in km
h_{te} is base station in meters
h_{re} is mobile antenna heights in meters
$a(h_m)$ is correction factor of mobile antenna height in dB

$$= (1.1 \log f_c - 0.7)h_{re} - (1.56 \log f_c - 0.8) \quad (4.19)$$
(for small cities)

$$= 8.29 (\log 1.54 h_{re})^2 - 1.1 \quad \text{for } f_c \geq 300 \text{ MHz} \quad (4.20)$$

$$= 3.2 (\log 11.75 h_{re})^2 - 4.97 \quad \text{for } f_c \leq 300 \text{ MHz} \quad (4.21)$$
(For large cities)

The figure 4.44 shows the M-file of Hata model, used to predict the path loss values from its equation.

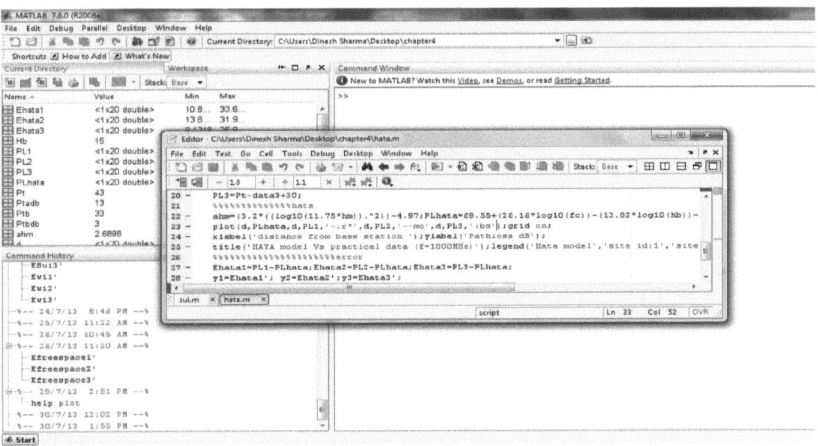

Figure 4.44 M-file of Hata Path Loss Model

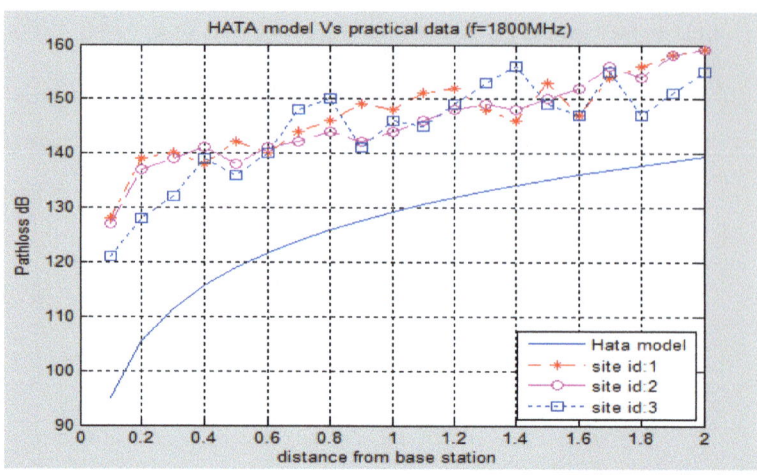

Figure 4.45 Comparison between Field Measured Path Loss and Hata Path Loss Model

From the figure 4.45 and 4.47, it is observed that the Hata model does not gives correct result in the terrain of Narnaul (Haryana). The error between the predicted value by the different path loss models and the measured path loss during test drive varies from 9 dB to 25 dB, which is a significant value. The Hata path loss does not provide any specific correction factor. The table 4.12 gives the variation of error between the predicted and the measured values. The figure 4.46 gives the information about the handoff limitation.

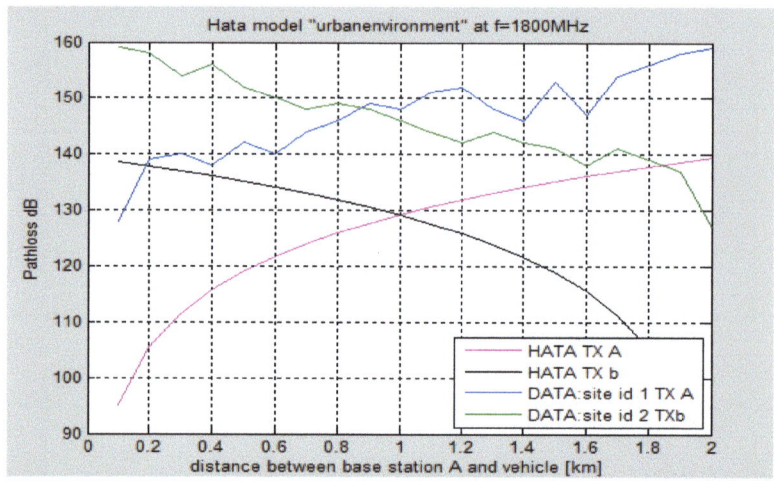

Figure 4.46 Variation of Path Loss between Hata Path Loss Model and Practical Field Data for Two Adjacent Cells

Table 4.12 Error between Measured and Hata Path Loss Model

Distance from base station (km)	Error (dB) Site id:NNL001	Error (dB) Site id:NNL002	Error (dB) Site id:NNL003
0.1	32.9008	31.9008	25.9008
0.2	33.6442	31.6442	22.6442
0.3	28.6446	27.6446	20.6446
0.4	22.3877	25.3877	23.3877
0.5	23.0858	19.0858	17.0858
0.6	18.3880	19.3880	18.3880
0.7	20.1070	18.1070	24.1070
0.8	20.1312	18.1312	24.1312
0.9	21.3883	14.3883	13.3883
1.0	18.8293	14.8293	16.8293
1.1	20.4190	15.4190	14.4190
1.2	20.1315	16.1315	17.1315
1.3	14.9471	15.9471	19.9471
1.4	11.8505	13.8505	21.8505
1.5	17.8296	14.8296	13.8296
1.6	10.8746	15.8746	10.8746
1.7	16.9776	18.9776	17.9776
1.8	18.1318	16.1318	9.1318
1.9	19.3318	19.3318	12.3318
2.0	19.5728	19.5728	15.5728

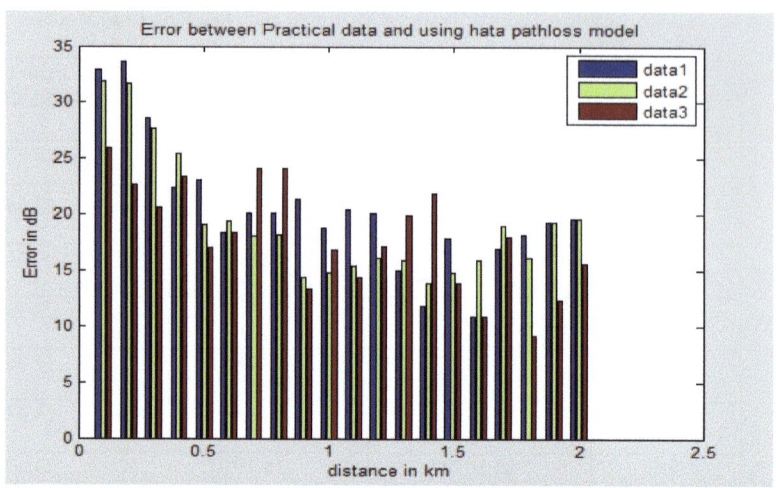

Figure 4.47 Variation of Error between Field Measured Data and SUI Path Loss Model

4.15 CONCLUSION

The different field propagation models has been compared with field measured data and plotted by using MATLAB. After examining the various field propagation models, is has been observed that no propagation model predict precisely for propagation terrain of Narnaul (Haryana). This is due to the fact that propagation of radio waves is strongly influenced by the nature of terrain particularly the size and density of buildings and climatic conditions [85]. Table 4.3-4.12 shows that error between measured path loss to the existing path loss predicted by different path loss models. The Okumura path loss model gives the best results in the terrain of Narnaul (South Haryana) because the error between the measured path loss and the predicted path loss by different path loss models is very less as depicted in figure 4.31 and substantiate table 4.8.

The model indicates a good agreement with the measured path loss. This is because Okumura model takes all the parameter into account such as free space loss, median attenuation relative to free space, transmitting/receiving antenna gain and gain due to environment, but this model is also does not give correct results. It means it requires some correction factor. The figure 4.30 shows the variation of path loss between the two adjacent cells which helps in deciding the handoff initiation. After analyzing the this figure it may be concluded that the distance of handoff initiation is same as measured by drive test and the obtained values by Okumura path loss model. So we can conclude that the Okumura model gives better results as compared to the other models. Therefore the Okumura path loss model gives better results in Narnaul (Haryana).

CHAPTER 5

THE EFFECT OF CLIMATIC CONDITIONS ON FIELD PROPAGATION MODEL

From the analysis carried out in previous chapter, it is observed that the Okumura path loss model is optimal model for Narnaul (Haryana) region. But still there is significant difference between currently measured field data and the predicted values by Okumura model. So to develop more precise model to get accurate results, the effects of atmosphere are discussed in this chapter.

5.1 INTRODUCTION

Radio signals are affected under atmospheric exposures, so in current research work the affect of climate is considered to include in existing propagation models. Mathematical modelling along with accurate characterization of radio channel is necessary to predict signal coverage and attain better data rates [182], [192]. Here it is considered an urban environment with different climatic conditions. Extreme climate conditions (heavy Fog, heavy snow, and heavy rain) influence radio communication links because the raindrops, cloud droplets, and solar flares geometrically scatter light. These particles affect the incoming radio signal in the channel entering at the receiver circuitry [60], [61]. In acute cases these variations can lead to complete cancellation of a signal at the receiving point. Some combinations of wind, temperature content and water content of the atmosphere cause attenuation of the radio signals [188]. These affects are minimum in the normal weather and noticeable in extreme weather conditions. Attenuation due to climate is generally proportionate to the frequency and wavelength of the radio wave [67], [161].

The related work which explains these affects is discussed in the further section.

5.1.1 Related Work

GSM signal strength measurements of two networks are taken during the dry season at meticulous points and the same measurements are again recorded during the rainy season. Losses of between 5dBm and 10dBm were observed in rural/suburban areas while the losses observed in the metropolis were between 0dB and 4dBm during heavy rain in rainy season. These losses are not much significant in urban areas, and are significant in the rural/suburban areas [170], [172]. Christian M. Ho performed an evaluation of radio wave propagation losses at SHF band by using available propagation models and modified the free space Friis Equation by adding atmospheric attenuation and fading effects [44]. The theoretical estimation and experimental investigation show that the lateral wave along air-canopy interface, the direct waves through the trunk layer and the canopy layer, and the ground-reflected waves are the main modes for propagation over large foliage depths at VHF band. The direct wave traveling through the canopy layer is the only wave that can be affected by the falling raindrops during a rain event [150], [151]. To analyze the climatic, wind and rain effects on a humid forested channel at VHF and UHF frequency bands, measurements are taken over a period of two weeks in a palm plantation located on the tropical island of Singapore during the northeast monsoon season (wettest month of the year). The experimental results indicate that the wind and rain can impose an additional attenuation on the electromagnetic signal propagating within the forest environment. The additional attenuation increases as the strength of the wind and rain increases [149]. The influence of the fog on the attenuation of the electromagnetic waves leads to disturbance in the wireless communication channel. In [40], it is explained that fog may be one of main factors in determination of the reliability of millimeter wave systems, where dense moist fog with high liquid water content happen frequently (coastal areas). Fog results from the concentration of atmospheric water vapor into water droplets that remain suspended in air [64], [69].

The climatic conditions of Narnaul (Haryana), INDIA are discussed here.

5.2 CLIMATIC CONDITIONS OF NARNAUL (HARYANA), INDIA

India is famous for its various climatic conditions and the weather conditions and climate can be divided into four principal subcategories. The climate of semi-desert region of India is very extreme. The maximum summer temperature reaches up to 50° Celsius and maximum winter temperature falls below 0° Celsius. The average annual rainfall is 20 cms. The table 5.1 and 5.2 depict the different seasons in India and various climatic regions of India respectively.

Table 5.1 Different Seasons of India

Name of the season	Tenure
Winter	The months of December to February
Summer	The months of March to june
Monsoon (rainy) season	The months of July to September
A post-monsoon period	The months of October and november

Table 5.2 Various Climatic Regions of India

Name of climatic region	States or territories
Tropical Rainforest	Assam and parts of the Sahyadri Mountain Range
Tropical Savannah	Sahyadri Mountain Range and parts of Maharashtra
Tropical and subtropical steppe	Parts of Punjab and Gujarat
Tropical Desert	Most parts of Rajasthan
Moist subtropical with winter	Parts of Punjab, Assam, and Rajasthan
Mountain climate	Parts of Jammu and Kashmir, Himachal Pradesh, and Uttaranchal
Drought	Rajasthan, Gujarat, and Haryana
Tropical semi-arid steppe	Tamil Nadu, Maharashtra, and other parts of South India

5.2.1 Geographical Location & Climate of NARNAUL (HARYANA)

Narnaul is located at $28^0 02'N$ $76^0 07'E$ 28.04^0 N $76.11^0 E$. The climatic condition of Narnaul is similar to the other parts of Haryana [88]. The long term analysis by Indian meteorological department (IMD) there is two climatic zones in Haryana (The north western part and the south western part of the state). The seasonal precipitation is negligible in many parts of Haryana.

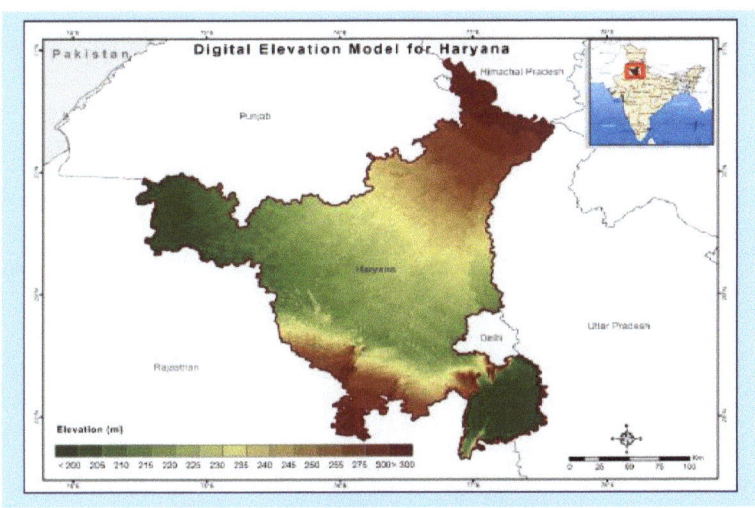

Figure 5.1 Geographical Map of Haryana

The climate of Haryana is very extreme i.e very hot (May and June) in summer and cold (December and January) in winters. Rainfall is varied, with the Shivalik Hills region being the wettest and the Aravali Hills region being the driest. The temperature falls to the lowest in winters and reaches upto 50° C during in summers. Winter months have average temperatures in the range 3°C to 9°C and the summer months temperatures are higher in the range of 48°C to 35°C.

Rainfall in the state of Haryana varies considerably both in space and time (from year to year). Table 5.3 gives the summary of observed rainfall statistics for Haryana. (Source: IMD rainfall data (2010-2012))

Table 5.3 Rainfall Statistics for Haryana

Season	Average value (mm)	Range (mm)
Annual rainfall	540	296.6- 1230.9
Winter	23.6	12-56
Monsoon	444.5	228-1010.3

The monsoon (June, July, August and September) rainfall (445 mm) contributes 80% of annual rainfall (540 mm). Inter annual variation in rainfall is very marginal and it is tabulated in table 5.4.

Table 5.4 Variation of Maximum and Minimum Temperature in Haryana

	Mean daily maximum temperature			
	Winter	Pre monsoon	monsoon	Post monsoon
Baseline	21	38	34	24
Mild century	22	40	36	25
End century	25	42	38	28
	Mean daily minimum temperature			
	Winter	Pre monsoon	monsoon	Post monsoon
Baseline	6	21	26	12
Mild century	7	23	28	14
End century	10	26	30	17

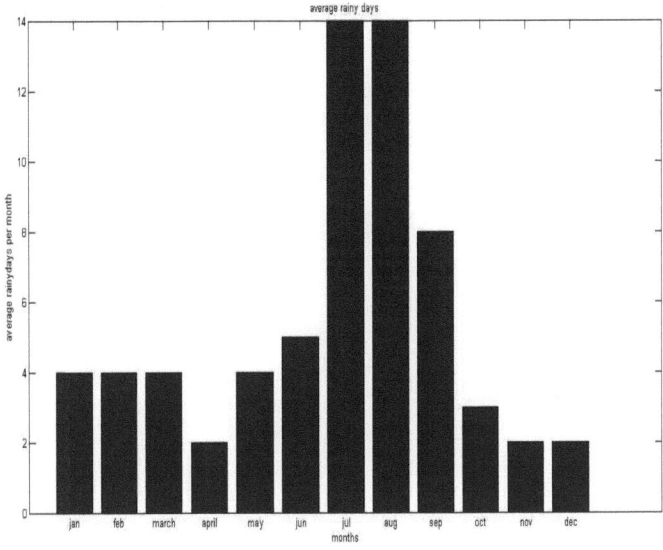

Figure 5.2 Average Rainy Days per Month in the Year 2011-2012

The figure 5.2 renders the average rainy days per month in the year 2011-2012 and figure 5.3 depicts the average rain fall statistics of the year 2011-2012.

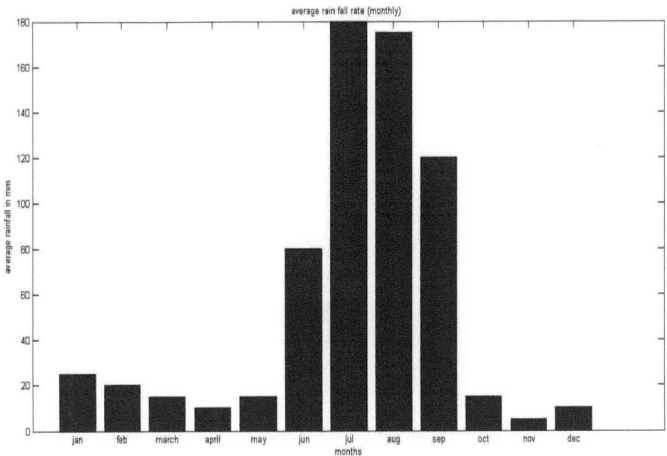

Figure 5.3 Average Rain Fall in Year 2011-2012

Fog is a natural phenomenon which has been observed during winter season in northern region of India. It has been described that fog forms from the condensation of atmospheric water vapor into water droplets that remain suspended in the air. The characterization of fog is based on water content, optical visibility, temperature and drop size distribution. Attenuation in foggy days may cause significant attenuation for radio relay links in climatic regions such as semi-desert terrain. The satellite view and climate during foggy day of Narnaul are shown in figure 5.4. The figure 5.5 depicts the average fog hours per day in 2011-2012.

Figure 5.4 Satellite View & Climate (foggy day) of Narnaul

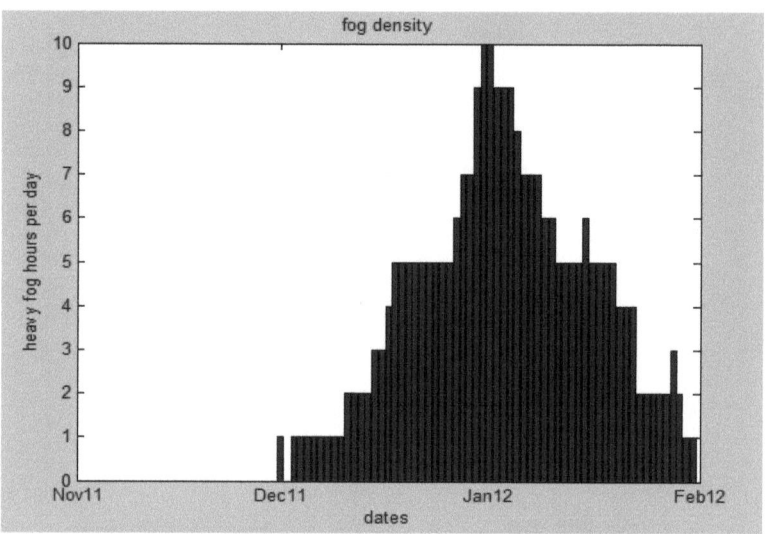

Figure 5.5 Average Fog Hours per Day in Year 2011-2012

5.3 COMPARSION & FIELD DATA COLLECTION DURING DIFFERENT CLIMATE CONDITIONS

As mentioned earlier the Okumura model gives more accurate results as compared to other path loss propagation models. Field data collection (Drive test) is the most common, easy and may be the best way to analyze network performance by means of coverage, system availability, network capacity and call quality. It gives idea about the downlink side of the radio propagation process, and also provides huge perspective to the service provider about what's happening with a subscriber point of view. The table 5.5 and table 5.6 depict the average received signal strength and path loss under different climatic conditions from base station having cell id NNL001.

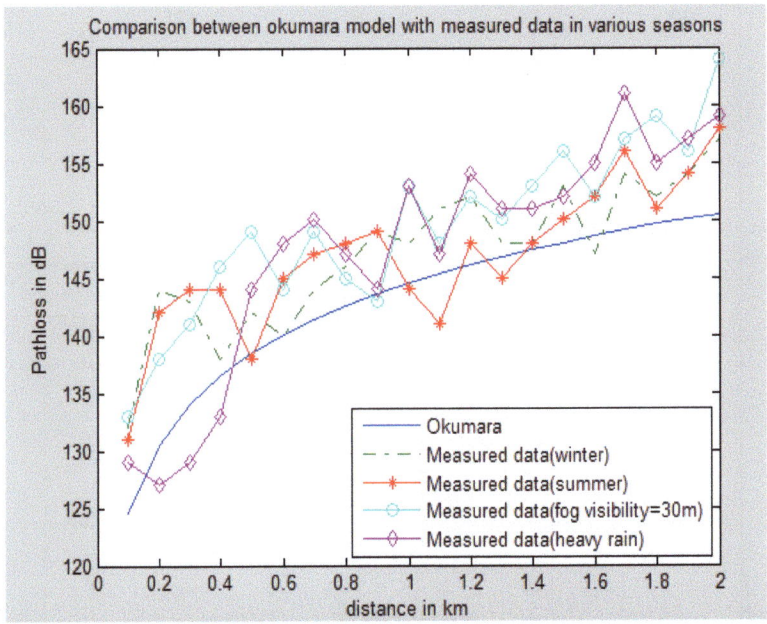

Figure 5.6 Variation of Path Loss in Different Climatic Conditions

Table 5.5 Average Signal Strength Measurements at Narnaul (Haryana)

Distance from base station (km)	Received signal strength (dBm) Summer	Received signal strength (dBm) Winter	Received signal strength (dBm) Heavy rain	Received signal strength (dBm) Heavy fog
0.1	-58	-59	-56	-60
0.2	-69	-71	-54	-65
0.3	-71	-70	-56	-68
0.4	-71	-65	-60	-73
0.5	-65	-69	-71	-76
0.6	-72	-67	-75	-71
0.7	-74	-71	-77	-76
0.8	-75	-73	-74	-72
0.9	-76	-76	-71	-70
1.0	-71	-75	-80	-80
1.1	-68	-78	-74	-75
1.2	-75	-79	-81	-79
1.3	-72	-75	-78	-77
1.4	-75	-75	-78	-80
1.5	-77	-80	-79	-83
1.6	-79	-74	-82	-79
1.7	-83	-81	-88	-84
1.8	-78	-79	-82	-86
1.9	-81	-81	-84	-83
2.0	-85	-84	-86	-91

Table 5.6 Average Path Loss Measurements at Narnaul (Haryana)

Distance from base station (km)	Path loss (dB) Summer	Path loss (dB) Winter	Path loss (dB) Heavy rain	Path loss (dB) Heavy fog
0.1	131	132	129	133
0.2	142	144	127	138
0.3	144	143	129	141
0.4	144	138	133	146
0.5	138	142	144	149
0.6	145	140	148	144
0.7	147	144	150	149
0.8	148	146	147	145
0.9	149	149	144	143
1.0	144	148	153	153
1.1	141	151	147	148
1.2	148	152	154	152
1.3	145	148	151	150
1.4	148	148	151	153
1.5	150	153	152	156
1.6	152	147	155	152
1.7	156	154	161	157
1.8	151	152	155	159
1.9	154	154	157	156
2.0	158	157	159	164

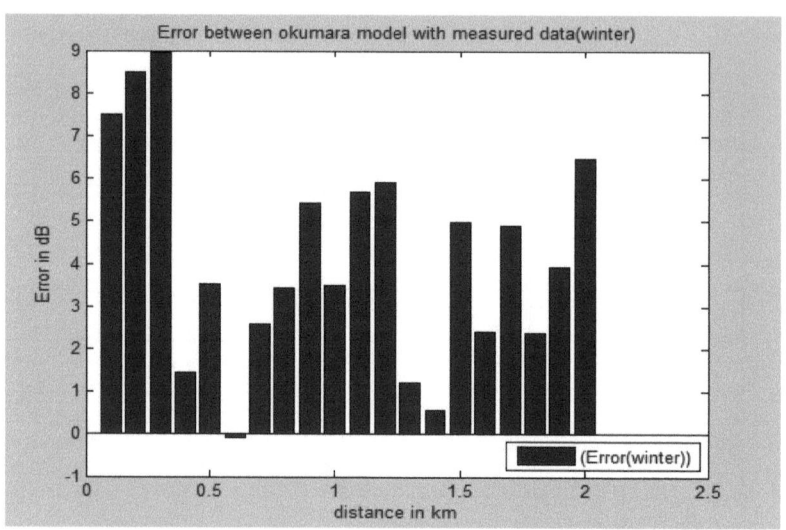

Figure 5.7 Error between Measured Data and Okumura Model in Winter

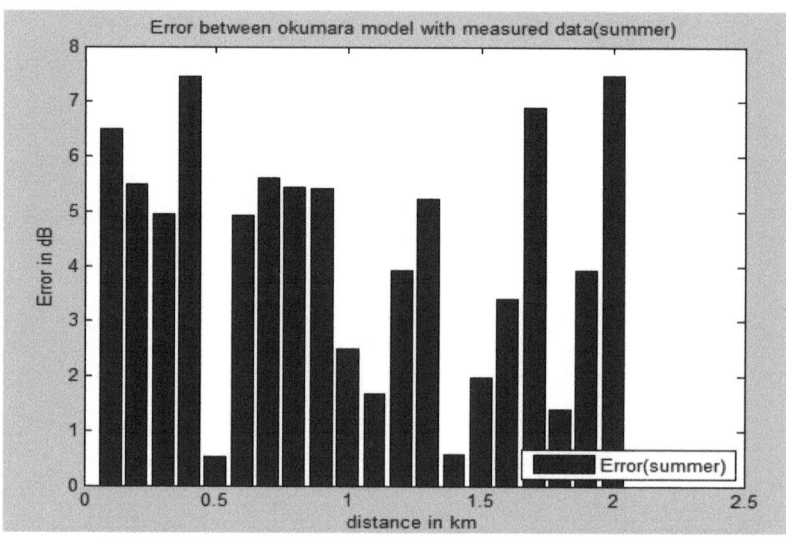

Figure 5.8 Error between Measured Data and Okumura Model in Summer

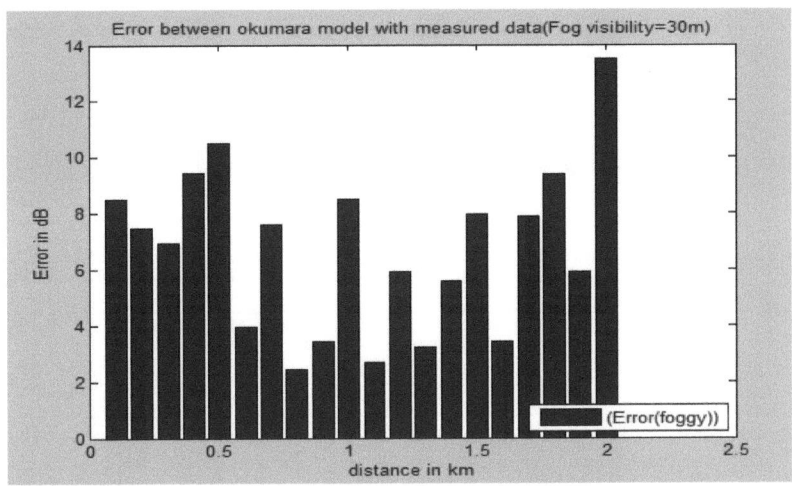

Figure 5.9 Error between Measured Data and Okumura Model in Heavy Fog Climate

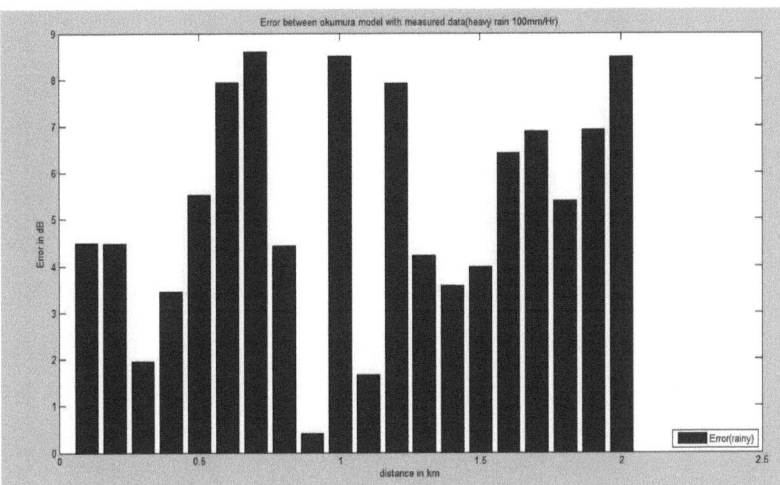

Figure 5.10 Error between Measured Data and Okumura Model in Heavy Rain Climate

    The figure 5.6 shows the variation of path loss in different climatic conditions such as winter, summer, heavy fog and heavy rain during drive test in Narnaul region. Figure 5.7, 5.8 , 5.9 and 5.10 respectively shows the variation of error between field measured data and path loss predicted by Okumura model in winter, summer, heavy fog and heavy rain climatic conditions. After observing the table 5.8, it is concluded that the error between currently measured field data (path loss) and the predicted path loss using Okumura model is more in the case of heavy fog environment and error is almost same in all other three climatic conditions (winter, summer and heavy rain).

Table 5.7 Error between Okumura Path Loss Model and Field Data

Distance from base station (km)	Error (dB) Summer	Error (dB) Winter	Error (dB) Heavy rain	Error (dB) Heavy fog
0.1	6.4964	7.4964	4.4964	8.4964
0.2	5.4758	8.4758	4.4758	7.4758
0.3	4.9540	8.9540	1.9540	6.9540
0.4	7.4552	1.4552	3.4552	9.4552
0.5	0.5170	3.5170	5.5170	10.5170
0.6	4.9334	-0.0666	7.9334	3.9334
0.7	5.5945	2.5945	8.5945	7.5945
0.8	5.4346	3.4346	4.4346	2.4346
0.9	5.4116	5.4116	0.4116	3.4116
1.0	2.4964	3.4964	8.4964	8.4964
1.1	1.6686	5.6686	1.6686	2.6686
1.2	3.9128	5.9128	7.9128	5.9128
1.3	5.2176	1.2176	4.2176	3.2176
1.4	0.5739	0.5739	3.5739	5.5739
1.5	1.9746	4.9746	3.9746	7.9746
1.6	3.4140	2.4140	6.4140	3.4140
1.7	6.8874	4.8874	6.8874	7.8874
1.8	1.3910	2.3910	5.3910	9.3910
1.9	3.9214	3.9214	6.9214	5.9214
2.0	7.4758	6.4758	8.4758	13.4758

5.4 DEVELOPMENT OF PROPAGATION PATH LOSS MODEL BY CONSIDERING DIFFERENT CLIMATIC CONDITIONS

The analysis of figure 5.6 to 5.10 shows that the best fit propagation path loss model (Okumura model) in the mentioned area does not follow the recorded data during drive test in extreme climatic conditions. This research mainly considers four different climatic conditions [183]. Generally at low frequencies the weather affects radio signal paths, especially below 50MHz, the ionosphere has a major effect, reflecting them back to Earth. At frequencies above 50MHz the troposphere has a major effect on the radio propagation path.

The interaction of radio signals with the ionized regions of the atmosphere makes radio propagation more complex to predict and analyze (analysis is simple in free space). A sudden ionosphere disturbance or shortwave fadeout is observed when the x-rays associated with a solar flare ionize the ionospheric D-region. These solar flares can disturb HF radio propagation and affect GPS accuracy.

5.4.1 Effect of Summer

Radio wave propagation is affected by the casual changes of water vapor in the troposphere due to the sun. Radiation from the Sun during large solar flares causes increased ionization in the D region which results in higher absorption of HF radio waves. Electrons are produced when solar radiation collides with uncharged atoms and molecules (Figure 5.11). Loss of free-electrons in the ionosphere occurs when a free electron combines with a charged ion to form a neutral particle (Figure 5.11). Loss of electrons occurs frequently, both day and night [100].

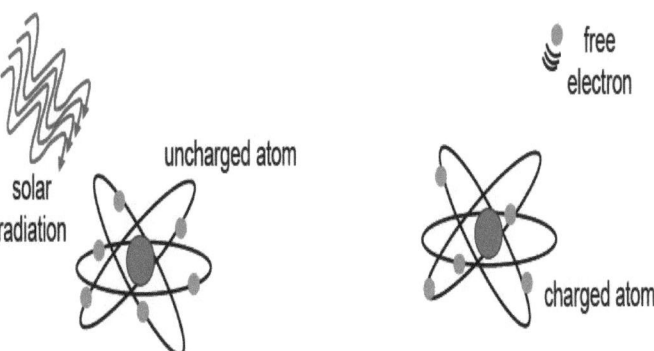

Figure 5.11 Illustration of Collision of Atoms and Molecules

The most important feature of the ionosphere in terms of radio communications is its ability to reflect radio waves. However, only those waves within a certain frequency range will be reflected. If the radio frequency is not too high, the pulses are reflected back towards the ground. The ionosonde records the time delay between transmission and reception of the pulses over a range of different frequencies. Sun effects are less at radio waves.

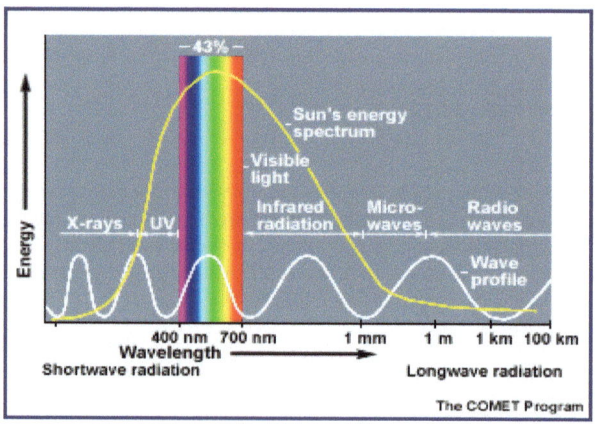

Figure 5.12 Effect of Sun in the Frequency Spectrum

From the figure 5.12 it is clear that the effect of sun is less in the case of radio waves.

5.4.2 Effect of Winter

Even the winter polar zone (a region of perpetual darkness) can suffer the effects of Polar Cap Absorption Events (PCA) since the ionised D region is formed by protons rather than sunlight. PCAs are attributed to high energy protons which escape from the Sun when a large flare occurs and move along the Earth's magnetic field lines to the Polar Regions. There they ionise the D region, causing very high absorption of HF waves. The effect of winter is less at radio frequencies.

5.4.3 Effect of Rain

In design of the radio links, the most desirable operating frequencies are below 10 GHz, because in such cases atmospheric absorption and rainfall loss may generally be neglected [44], [66], [78]. In rainy season the increased earth moisture content gives rise to greater ground conductivity. The ordinary rain i.e. less than 100 mm does not affect the GSM signal strength but the heavy rain fall i.e. greater than 100mm affects the GSM signal strength [13], [138].

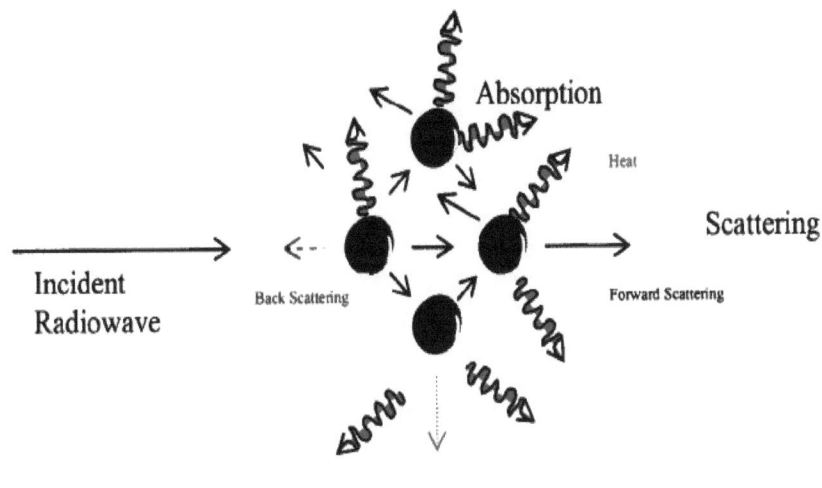

Figure 5.13 Effect of Rain on Radio Waves

Détermination of rain atténuation by ITU Radio communication Assembly, The specific attenuation (dB/km) is obtained from the rain rate R (mm/h) using the power law relationship [51]:

$$Sa_R = K*(R^L) \tag{5.1}$$

Values for the coefficients K and L are determined as functions of frequency, f (GHz)

$$K = [K_H + K_V + (K_H - K_V)\cos^2\phi \cos 2\varphi]/2 \tag{5.2}$$

$$L = [K_H L_H + K_V L_V + (K_H L_H - K_V L_V)\cos^2\phi \cos 2\varphi]/2K \tag{5.3}$$

Where,

$\phi=20$; path elevation angle

$\varphi=45$; polarization tilt angle

K_H=0.0000678; K_V=0.0000874;

L_H=1.0564; L_V=0.9102;

$Rain attenuation = RA = Sa_R * d$ in dB \hspace{1cm} (5.4)

5.4.4 Effect of Fog

Fog is a natural phenomenon which has been observed during winter season in this region of India. It is described that fog forms from the condensation of atmospheric water vapor into water droplets that remain suspended in the air. It is generally difficult to directly measure fog density or to obtain its statistical data. The characterization of fog is based on water content, optical visibility, temperature and drop size distribution [71]. There are mainly 4 types of fog exist in some types of environments. Those are strong advection fog, light advection fog, strong radiation fog, and light radiation fog [109]. Attenuation in the foggy days may cause significant attenuation for radio waves in extreme climatic regions such as semi-desert terrain. The effect of the attenuation due to fog, fog attenuation (dB/km), can be related to atmosphere visibility, V (km), defined by the maximum distance that we can recognize a black object against the sky [217]. The relation between fog attenuation and visibility are expressed by the following equation [217].

$$\text{Fog attenuation (FA)} = 10 \log_{10}(\varepsilon)/V \quad \text{in dB/km} \tag{5.5}$$

On the basis of the field data collection and the attenuation by rain and fog the Okumura path loss model can be developed as

$$\text{Developed Okumura model} = \text{Okumura model} + AF + CF \tag{5.6}$$

Where, AF is area factor it can vary according to area and the formula of area factor considered here is

Area factor = mean value (Difference between Path loss predicted by Okumura model and field measured data of path loss considering all seasons)

And CF is climate factor, it is zero for normal weather conditions, equal to fog attenuation for heavy fog conditions and equal to rain attenuation for heavy rain conditions.

$$CF = \begin{cases} 0 & \text{(For ordinary climate conditions)} \\ 10 \log_{10}(\varepsilon)/V * d & \text{(For dense fog less than 50m)} \\ Sa_R * d & \text{(Heavy rain greater than are equal to 100mm)} \end{cases} \tag{5.7}$$

The equation 5.7 gives the value of climatic factor.

5.5 COMPARATIVE ANALYSIS OF FIELD MEASURED DATA, OKUMURA MODEL AND DEVELOPED OKUMURA MODEL

The currently taken field data is measured at 1800 MHz, Narnaul city (Haryana, India). The measurements were taken at regular intervals by considering twenty points in the circle of 2 km from the BTS. With the help of the above analysis it is found that path loss model Okumura gives results nearer to the practical data but not accurate results. The analysis shows that the measured data is nearer to the Okumura model and still need some modification to get more accurate results. The modification of Okumura model is given by considering climatic effects along with the area factor in the above mentioned area where drive test is conducted.

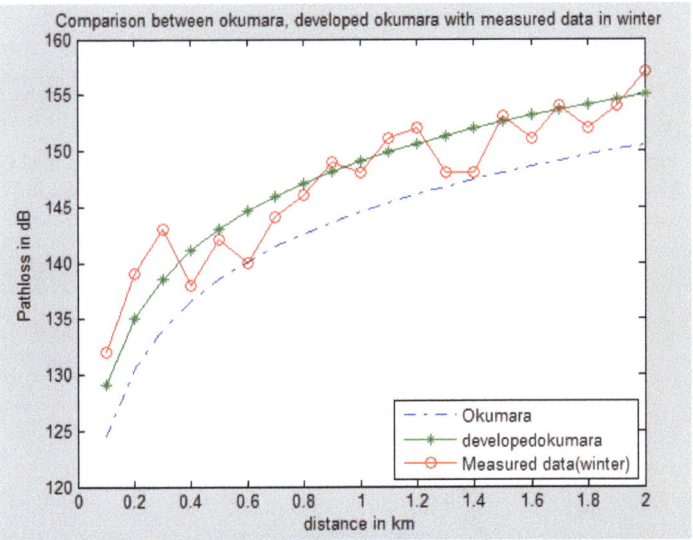

Figure 5.14 Comparison between Field Measured Data, Okumura and Developed Okumura Path Loss Model in Winter Climate

The figure 5.14 shows comparison between field measured data, Okumura and developed Okumura pathloss model. The field data was taken in an ordinary winter climate. the winter data is approximated using a fourth degree polynomial equation then it is compared with okumura and developed okumura model with the help of matlab in figure 5.15. The approximated fourth degree polynomial equation is (field data in terms of diatance)

$$Fwinter(d) = 4.2689\ d^4 - 13.858\ d^3 + 8.1317\ d^2 + 16.631\ d + 133.05 \qquad (5.7)$$

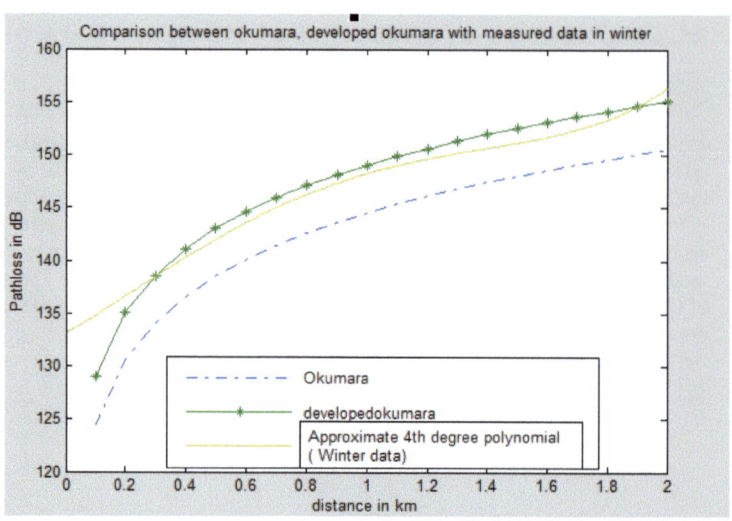

Figure 5.15 Comparison between Approximated 4th Degree Polynomial Curve of Field Measured data, Okumura and Developed Okumura Path Loss Model in Winter Climate

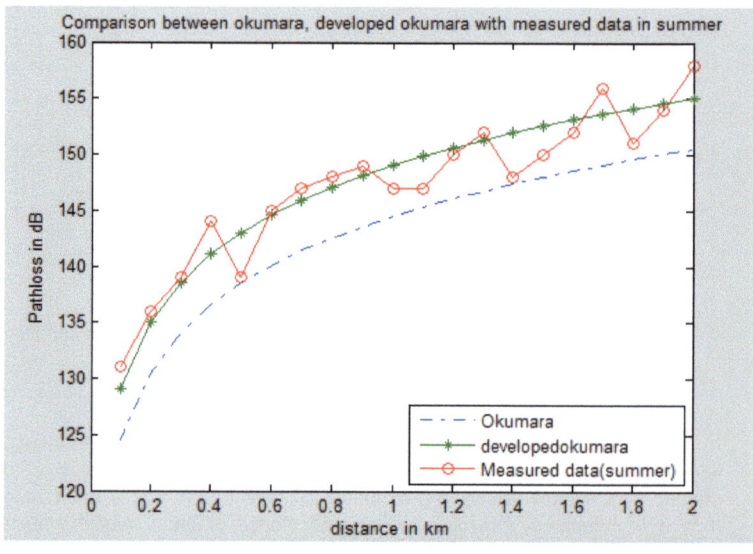

Figure 5.16 Comparison between Field Measured Data, Okumura and Developed Okumura Path Loss Model in Summer Climate

The figure 5.16 shows comparison between field measured data okumura and developed okumura pathloss model. The field data was taken at ordinary summer climate. the summer data is approximated using a fourth degree polynomial equation and plotted that equation and then compared with okumura and developed okumura model using matlab in figure 5.17. The approximated fourth degree polynomial equation is:

$$F_{summer}(d) = -1.6896\, d^4 - 16.929\, d^3 - 45.393\, d^2 + 51.615\, d + 126.86 \qquad (5.8)$$

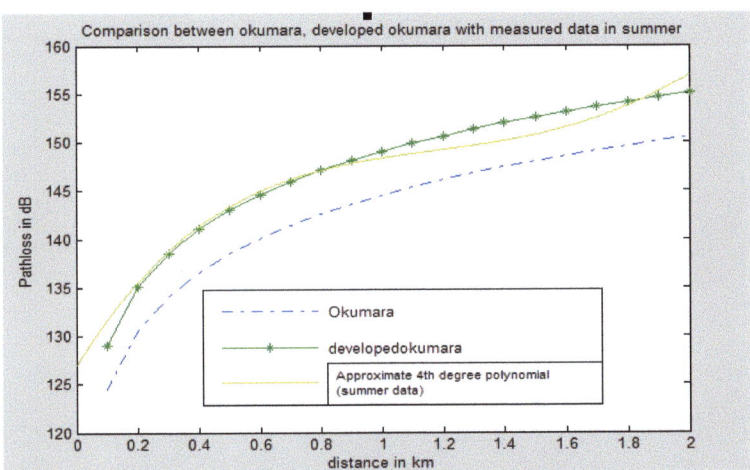

Figure 5.17 Comparison between Approximated 4th Degree Polynomial Curve of Field Measured Data, Okumura and Developed Okumura Path Loss Model in Summer Climate

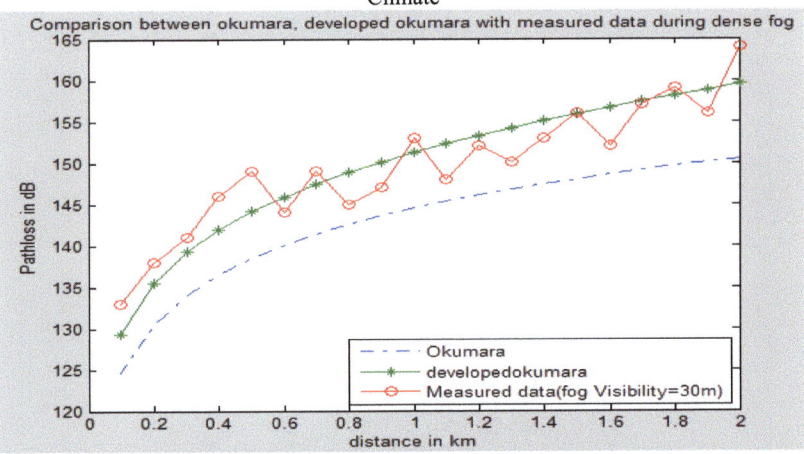

Figure 5.18 Comparison between Field Measured Data, Okumura and Developed Okumura Path Loss Model in Heavy Fog Climate

The figure 5.18 shows comparison between field measured data okumura and developed okumura pathloss model. The field data was taken at heavy fog condition (visibility 30m). the fog data is approximated using a fourth degree polynomial equation and plotted that equation and then compared with okumura and developed okumura model using matlab in figure 5.19. the approximated fourth degree polynomial equation is:

$$Ffog(d) = -8.3008\, d^4 + 46.218\, d^3 - 85.291\, d^2 + 69.277\, d + 127.33 \tag{5.9}$$

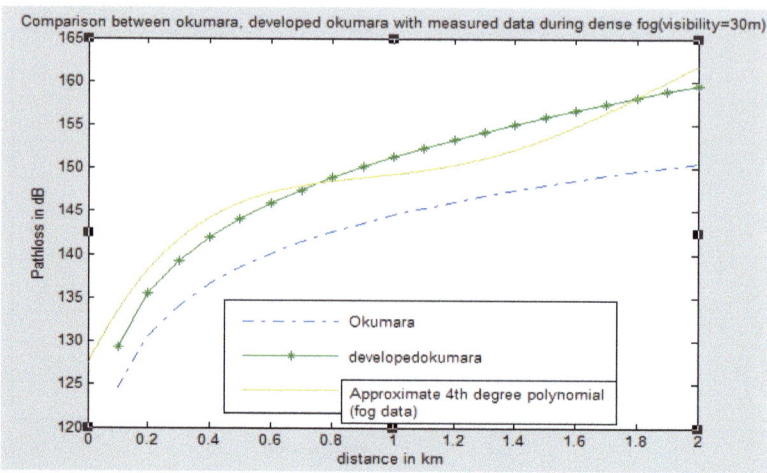

Figure 5.19 Comparison between Approximated 4th Degree Polynomial Curve of Field Measured Data, Okumura and Developed Okumura Path Loss Model in Heavy Fog Climate

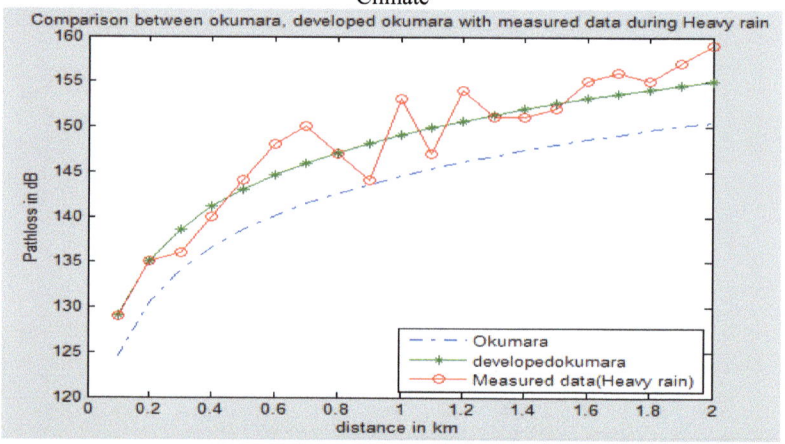

Figure 5.20 Comparison between Field Measured Data, Okumura and Developed Okumura Path Loss Model in Heavy Rain Climate

The figure 5.20 shows comparison between field measured data okumura and developed okumura pathloss model. The field data was taken at heavy raincondition (100mm/Hr). the rain data is approximated using a fourth degree polynomial equation and plotted that equation and then compared with okumura and developed okumura model using matlab in figure 5.21. the approximated fourth degree polynomial equation is:

$$\text{Frain}(d) = -5.434\, d^4 + 33.955\, d^3 - 72.905\, d^2 + 71.683\, d + 122.26 \tag{5.10}$$

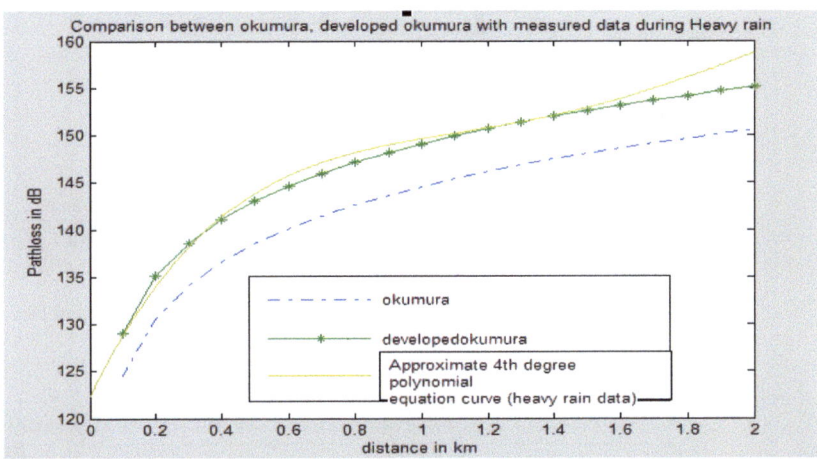

Figure 5.21 Comparison between Approximated 4th Degree Polynomial Curve of Field Measured Data, Okumura and Developed Okumura Path Loss Model in Heavy Rain Climate

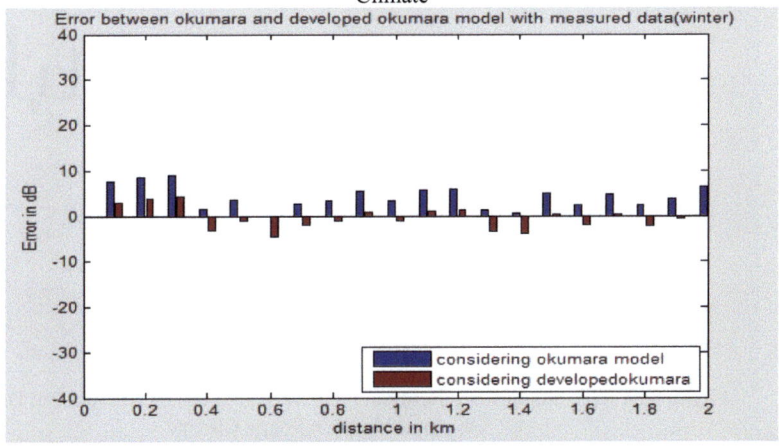

Figure 5.22 Variation of Error between Field Measured Data, Okumura and Developed Okumura Model in Winter

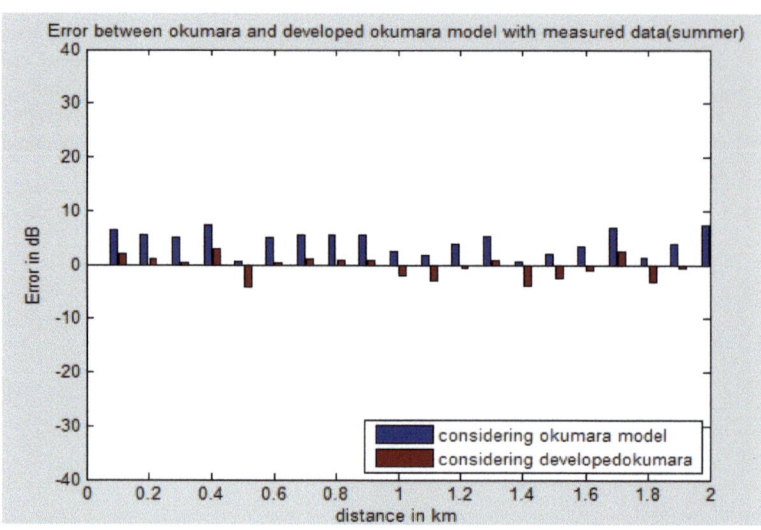

Figure 5.23 Variation of Error between Field Measured Data, Okumura and Developed Okumura Model in Summer

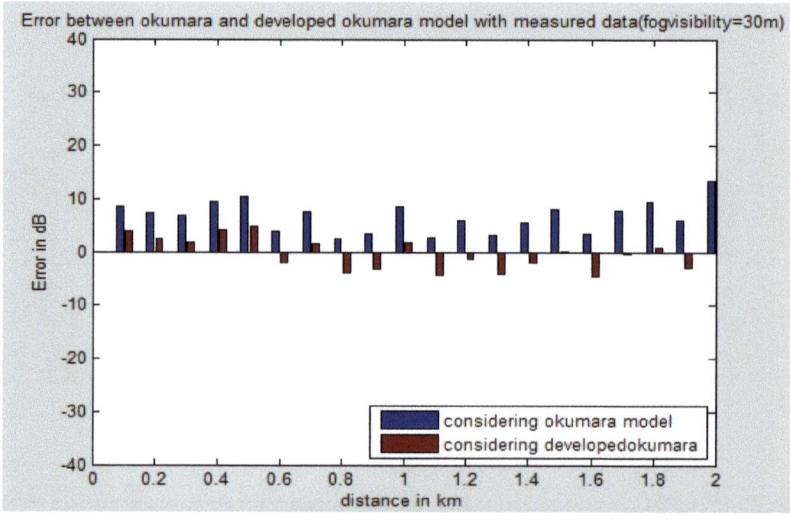

Figure 5.24 Variation of Error Between Field Measured Data, Okumura and Developed Okumura Model in Foggy Climate

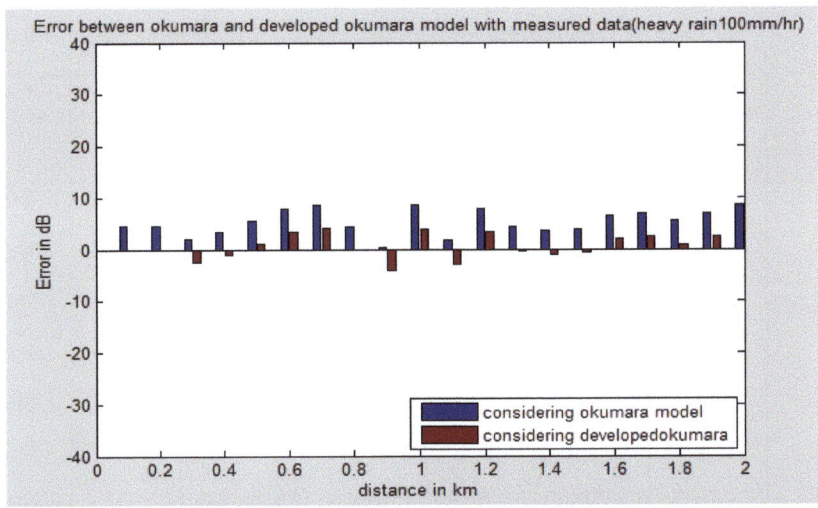

Figure 5.25 Variation of Error between Field Measured Data, Okumura and Developed Okumura Model in Winter

The figures 5.22, 5.23 & 5.24 and the table 5.8 shows the variation of error between field measured data, okumura and developed okumura path loss model. The error between field measured data and developed okumura model is less compared with okumura model. Hence it can be concluded that the developed okumura model prediction is better than the original okumura model in Narnaul city (Haryana, INDIA).

5.6 VALIDATION OF DEVELOPED OKUMURA PATH LOSS MODEL

The developed okumura model is given as:

Developed Okumura model= Okumura model+AF+CF

Where, AF is area factor

CF is climatic factor

The area factor is correction factor and it is included in the formula to consider area wise effects and CF included in the formula to consider variations in seasons (winter, winter with dense fog, summer, rainy season) which affect mobile radio propagation.

Table 5.8 Error between Measured, Okumura and Developed Okumura Model

Distance from base station (km)	Error (dB) summer	Error (dB) Winter	Error (dB) Heavy rain	Error (dB) Heavy fog
0.1	1.9828	2.9828	-0.0179	3.7609
0.2	0.9622	3.9622	-0.0392	2.5185
0.3	0.4403	4.4403	-2.5617	1.7748
0.4	2.9416	-3.0584	-1.0612	4.0542
0.5	-3.9966	-0.9966	1.0000	4.8941
0.6	0.4197	-4.5803	3.4157	-1.9114
0.7	1.0808	-1.9192	4.0761	1.5279
0.8	0.9210	-1.0790	-0.0845	-3.8538
0.9	0.8979	0.8979	-4.1082	-3.0987
1.0	-2.0172	-1.0172	3.9760	1.7643
1.1	-2.8451	1.1549	-2.8526	-4.2854
1.2	-0.6009	1.3991	3.3910	-1.2631
1.3	0.7039	-3.2961	-0.3049	-4.1802
1.4	-3.9398	-3.9398	-0.9493	-2.0457
1.5	-2.5391	0.4609	-0.5492	0.1332
1.6	-1.0996	-2.0996	1.8895	-4.6492
1.7	2.3738	0.3738	2.3623	-0.3977
1.8	-3.1227	-2.1227	0.8651	0.8840
1.9	-0.5923	-0.5923	2.3948	-2.8075
2.0	2.9622	1.9622	3.9486	4.5252

The validation of developed Okumura propagation path loss model comprises two approaches:

> By taking reference model
> By applying the developed model in another city (similar to Narnaul)

5.6.1 By Taking Reference Model

In this approach two reference models has been considered, one is for fog attenuation and another is for rain attenuation to validate developed Okumura propagation path loss model.

5.6.1.1 Fog Attenuation Reference Model

In 1984 Altshuler gave a formula for fog attenuation which depends on wavelength and temperature. The equation for attenuation in $[(dB/km)/(g/m^3)]$ is given as[202]:-

$$A = -1.347 + 0.0372\lambda + 18/\lambda - 0.022 T_0 \qquad (5.11)$$

here, λ is wavelength in mm
T_0 temperature ^0C

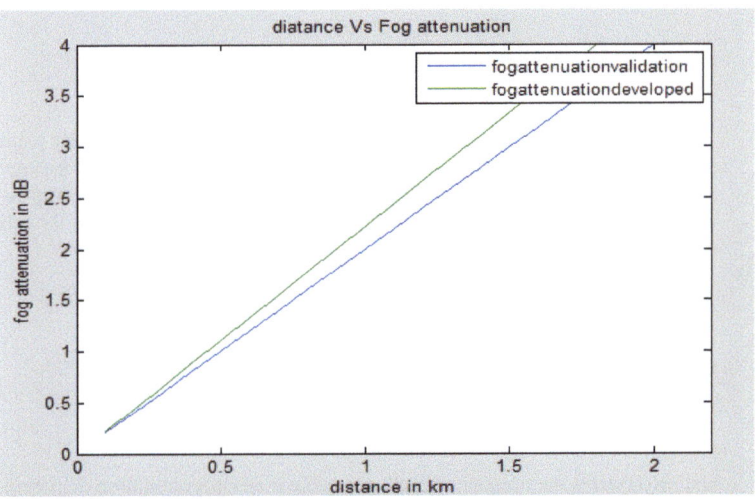

Figure 5.26 Comparison between Developed Fog Attenuation and Reference Fog Attenuation Model.

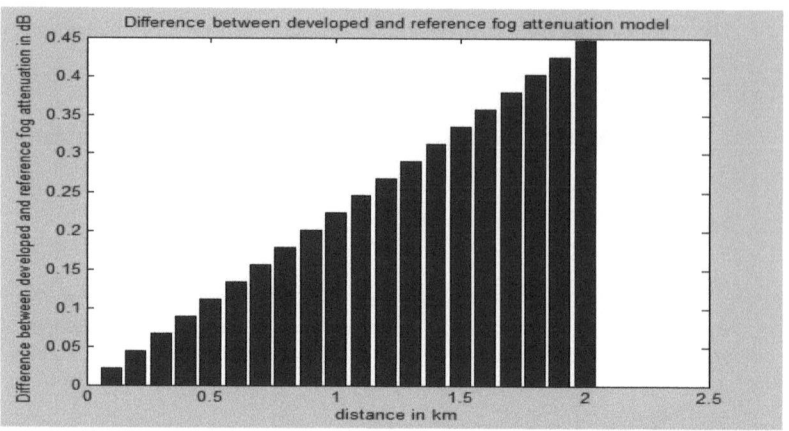

Figure 5.27 Difference between Developed and Reference Fog Attenuation Model

The figure 5.2 shows the comparsion between developed okumura model and reference model for fog attenuation model. The figure 5.27 and table 5.9 shows the error between fog attenuation of developed model and the reference model. The maximum error between these two models is 0.45 which is very less. Hence it has been conclude that the error between the reference model and the fog attenuation part of developed model is very less. Therefore the fog attenuation part of developed model predict correct values of signal strength.

5.6.1.2 Rain Attenuation Reference Model

M. Sridhar gave an equation to calculate rain attenuation. He explained that the rain drops absorb most of the electromagnetic energy and some of the energy gets scattered by Rayleigh and Mie scattering mechanisms [214]. The rain drop size distribution is exponential and the mathematical expression is given in the following equation 5.12.

$$N(D) = N_0 e^{(\frac{-D}{D_m})} \text{ m}^{-3} \tag{5.12}$$

Here, D_m is median rain drop diameter

$$D_m = 0.122 \times R^{0.21} \text{ dB/km} \tag{5.13}$$

and N(D)dD is the number of drops per cubic meter with rain drop diameters between D and D + dD mm [140]. The rainfall rate R in terms of N (D) and terminal velocity of V (D) is given as:

$$R = 0.6 \times 10^{-3} \pi \int D^3 V(D) N(D) dD \tag{5.14}$$

The expression for rain attenuation is given below:

$$A = \int \left(\frac{D^3}{\lambda}\right)\left(N_0 \times e^{\left(\frac{-D}{D_m}\right)}\right) \quad \text{dB/km} \tag{5.15}$$

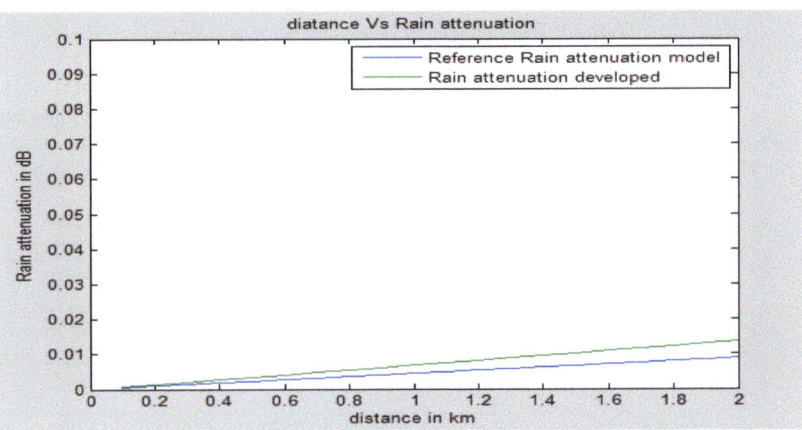

Figure 5.28 Comparison between Developed Rain Attenuation and Reference Rain Attenuation Model

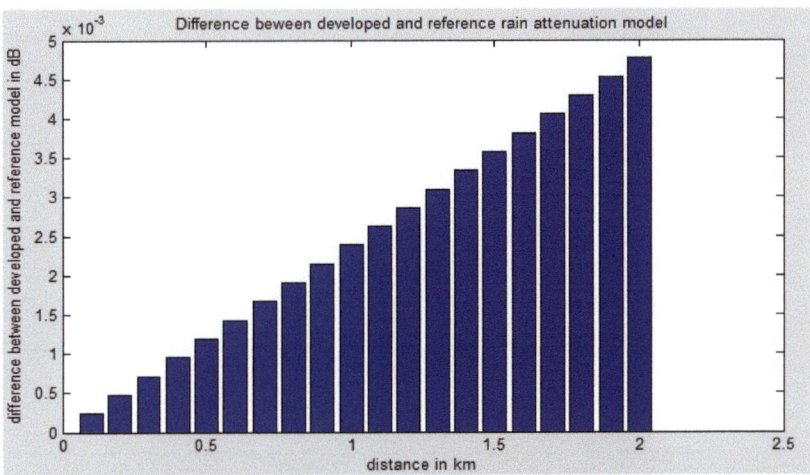

Figure 5.29 Difference between Developed and Reference Rain Attenuation

The figure 5.28 depicts the comparison between developed rain attenuation factor given in developed model and reference rain attenuation factor. After examining, it is observed that the developed attenuation is giving results same as that of reference attenuation factor taken for validation. The figure 5.29 and table 5.10 shows the difference between developed and reference rain attenuation. The maximum difference between these two attenuations is approximately zero (0.0048). Hence it has been concluded that the prediction made by developed model and reference model are same.

Table 5.9 Error between Fog Attenuation and Reference Model

Distance from base station (km)	Fog attenuation Considered in the modified model	Fog attenuation reference model	Difference between both fog attenuation equations
0.1	0.2219	0.1995	0.0223
0.2	0.4437	0.3991	0.0446
0.3	0.6656	0.5986	0.0670
0.4	0.8874	0.7981	0.0893
0.5	1.1093	0.9976	0.1116
0.6	1.3311	1.1972	0.1339
0.7	1.5530	1.3967	0.1562
0.8	1.7748	1.5962	0.1786
0.9	1.9967	1.7958	0.2009
1.0	2.2185	1.9953	0.2232
1.1	2.4404	2.1948	0.2455
1.2	2.6622	2.3944	0.2678
1.3	2.8840	2.5939	0.2902
1.4	3.1059	2.7934	0.3125
1.5	3.3278	2.9929	0.3348
1.6	3.5496	3.1925	0.3571
1.7	3.7715	3.3920	0.3795
1.8	3.9933	3.5915	0.4018
1.9	4.2152	3.7911	0.4241
2.0	4.4370	3.9906	0.4464

Table 5.10 Error between Rain Attenuation and Reference Model

Distance from base station (km)	Rain attenuation Considered in the modified model	Rain attenuation reference model	Difference between both Rain attenuation equations
0.1	0.0007	0.0004	0.0002
0.2	0.0014	0.0009	0.0005
0.3	0.0020	0.0013	0.0007
0.4	0.0027	0.0018	0.0010
0.5	0.0034	0.0022	0.0012
0.6	0.0041	0.0026	0.0014
0.7	0.0047	0.0031	0.0017
0.8	0.0054	0.0035	0.0019
0.9	0.0061	0.0040	0.0021
1.0	0.0068	0.0044	0.0024
1.1	0.0075	0.0048	0.0026
1.2	0.0081	0.0053	0.0029
1.3	0.0088	0.0057	0.0031
1.4	0.0095	0.0062	0.0033
1.5	0.0102	0.0066	0.0036
1.6	0.0109	0.0070	0.0038
1.7	0.0115	0.0075	0.0041
1.8	0.0122	0.0079	0.0043
1.9	0.0129	0.0084	0.0045
2.0	0.0136	0.0088	0.0048

5.6.2 By Applying the Developed Model in another City

In this approach, the validation of developed propagation path loss model has been applied at different city of Haryana, India having the same climatic conditions. Here the city Hisar, Haryana has been taken whose long. 75.7 and lat. 29.1. The derive test has been performed in Hisar by considering a BTS id in highly populated area. The received signal strength in region of hisar has been shown in table 3.7-3.10 in different months or in different climatic conditions.

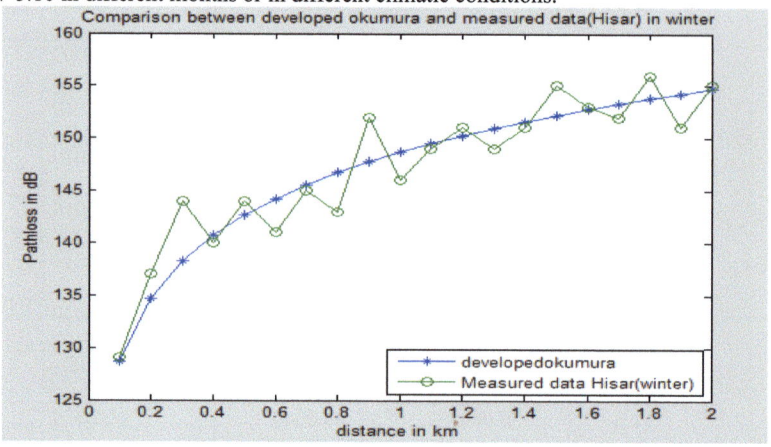

Figure 5.30 Comparison between Developed Okumura Model and Field Data Taken (Hisar, Haryana, INDIA) in Winter Season

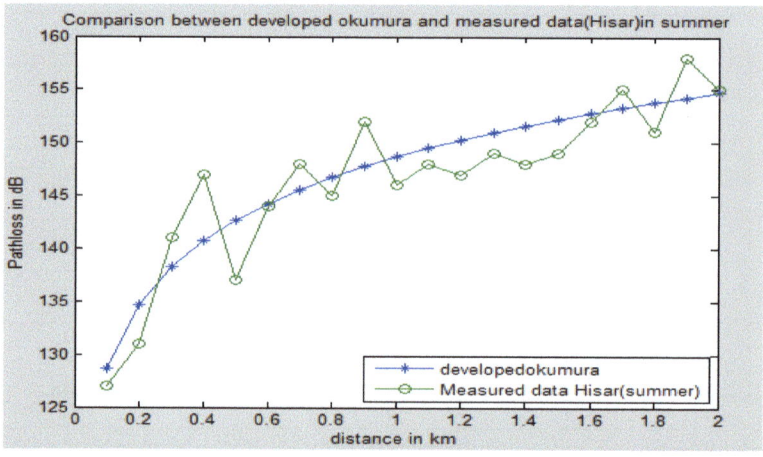

Figure 5.31 Comparison between Developed Okumura Model and Field Data Taken (Hisar, Haryana, INDIA) in Summer Season

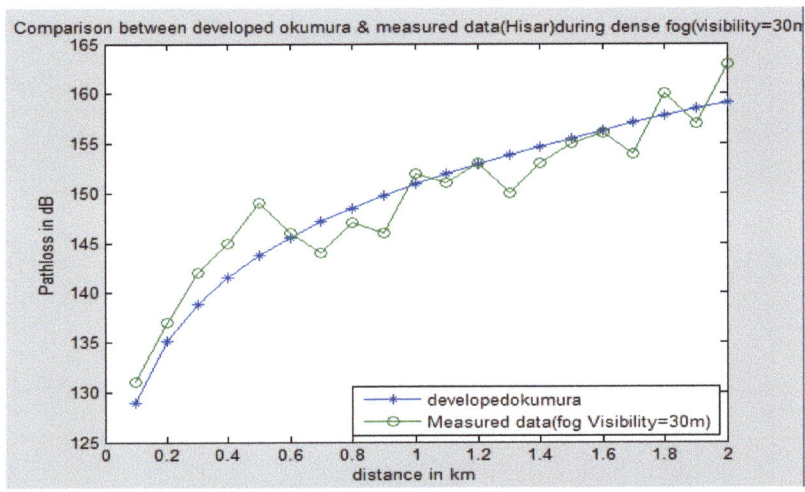

Figure 5.32 Comparison between Developed Oumura Model and Field Data Taken (Hisar, Haryana, INDIA) in Heavy Fog Condition

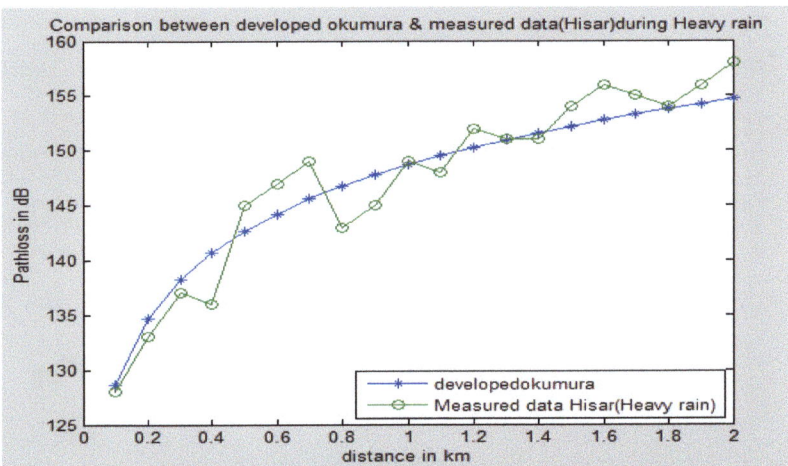

Figure 5.33 Comparison between Developed Okumura Model and Field Data Taken (Hisar, Haryana, INDIA) in Heavy Rain Condition

The figures 5.30-5.33 shows the comparsion between the field measured data in different climatic conditions of Hisar and developed okumura propagation path loss model. The field measured data has been taken in Hisar in diferent climatic conditions, winter, summer, rain and fog. In these four climatic conditions, a comparative analysis has been done. The figures 5.34-5.37 shows the error between the developed okumura model and the field measured data. The table 5.11 shows the exact between the developed okumura model and the field measured data. After careful examination of table 5.11, it has been observed the error between the predicted values by developed model and the field measured data is less. Hence it has been conclude that the developed okumura propagation path loss model predict more accurately.

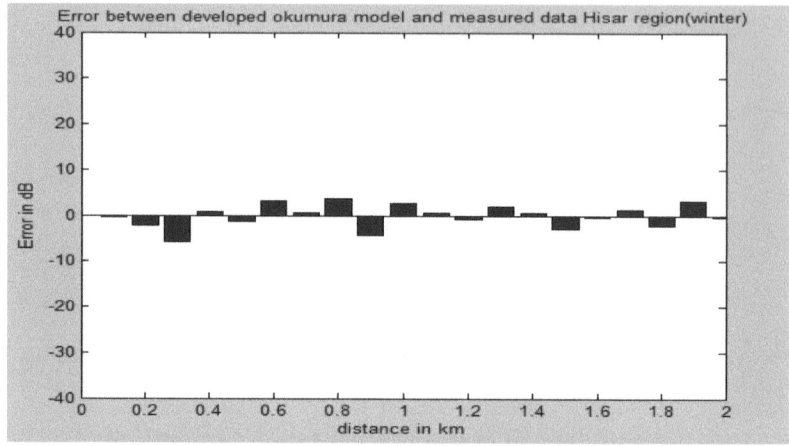

Figure 5.34 Error between Developed Okumura Model and Field Data Taken (Hisar, Haryana, INDIA) in Winter Condition

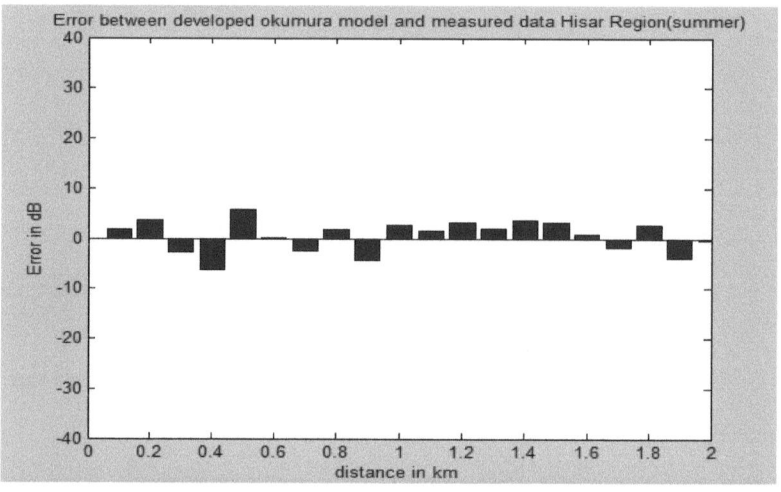

Figure 5.35 Error between Developed Okumura Model and Field Data taken (Hisar, Haryana, INDIA) in Summer Season

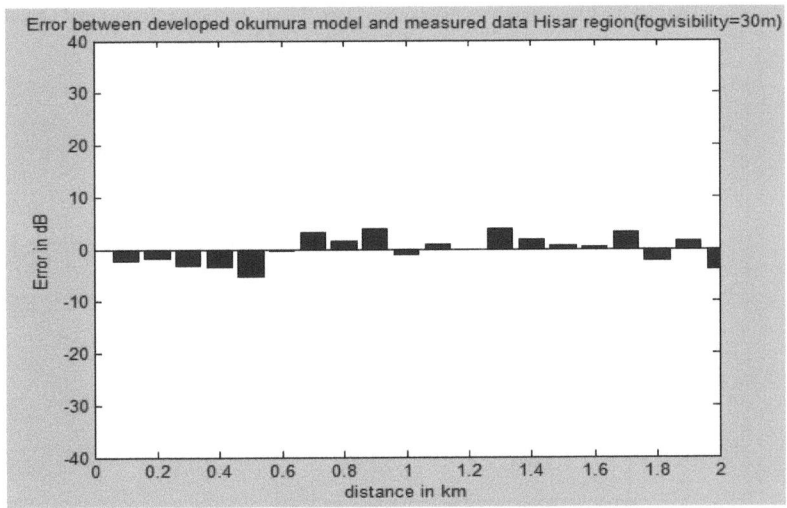

Figure 5.36 Error between Developed Okumura Model and Field Data Taken (Hisar, Haryana, INDIA) in Heavy Fog Condition

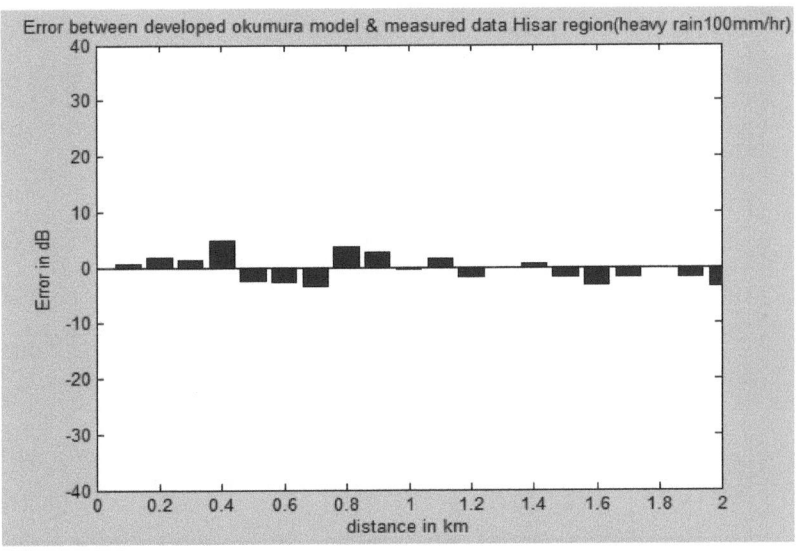

Figure 5.37 Comparison between Developed Okumura Model and Field Data Taken (Hisar, Haryana, INDIA) in Heavy Rain Condition

Table 5.11 Error between Measured Data (Hisar, Haryana, India) and Developed Okumura Model

Distance from base station (km)	Error (dB) Winter	Error (dB) Summer	Error (dB) Heavy fog	Error (dB) Heavy rain
0.1	-0.3328	1.6672	-2.1109	0.6679
0.2	-2.3122	3.6878	-1.8685	1.6892
0.3	-5.7903	-2.7903	-3.1248	1.2117
0.4	0.7084	-6.2916	-3.4042	4.7112
0.5	-1.3534	5.6466	-5.2441	-2.3500
0.6	3.2303	0.2303	-0.4386	-2.7657
0.7	0.5692	-2.4308	3.1221	-3.4261
0.8	3.7290	1.7290	1.5038	3.7345
0.9	-4.2479	-4.2479	3.7487	2.7582
1.0	2.6672	2.6672	-1.1143	-0.3260
1.1	0.4951	1.4951	0.9354	1.5026
1.2	-0.7491	3.2509	-0.0869	-1.7410
1.3	1.9461	1.9461	3.8302	-0.0451
1.4	0.5898	3.5898	1.6957	0.5993
1.5	-2.8109	3.1891	0.5168	-1.8008
1.6	-0.2504	0.7496	0.2992	-3.2395
1.7	1.2762	-1.7238	3.0477	-1.7123
1.8	-2.2273	2.7727	-2.2340	-0.2151
1.9	3.2423	-3.7577	1.4575	-1.7448
2.0	-0.3122	-0.3122	-3.8752	-3.2986

5.7 CONCLUSION

In this chapter the field measured data in different climatic conditions has been compared with the okumura path loss model. After the analysis, it has been observed that the error between measured value and predicted is an optimal value. On the basis of this analysis a pathloss prediction model has been proposed. The developed okumura pathloss model contains area factor and climate factor along with the original okumura model. It has been discussed that the pathloss models are designed on the basis of the area and environmental conditions. With the help of developed okumura model , the pathloss prediction is more precise as compared to the original okumura model.

Validation of developed okumura model has been taken using the reference attenuation models for rain and fog. The reference rain attenuation equation is taken from M. Sridhar research paper and the reference fog attenuation is taken from Altshuler research paper. In this investigation the difference in both attenuation equations is very small. Further the developed okumura model is compared with field measured data taken at Hisar (Haryana, India). This investigation shows that the developed model is valid through out the Haryana and the area which has same climatic conditions similar to Narnaul.

CHAPTER 6

CELL COVERAGE AREA AND EFFECT ON LINK BUDGET DUE TO CLIMATIC CONDITIONS

The network planning is a significant step in establishment of wireless communication system. In this chapter the concentration is completely on pre-process of network planning i.e. link budgeting of GSM cellular system. Now, the effects of climatic conditions on link budget and coverage area made in present work are discussed.

6.1 INTRODUCTION

Radio planning for networks is an important part of network development. Insufficient network planning results in over design and waste of resources under design and poor system performance. Link budget is necessary before designing a radio communication system. The link budget enables factors like required antennas gain levels, radio transmitter power levels, and receiver sensitivity figures. Link budget is often used for satellite systems. In any communication system it is essential that the required signal levels are maintained to ensure that the received signal levels are sufficient and above the noise level. To maintain the signal levels in a wireless communication system, larger antennas and high transmitter power levels are required and those can be added considerably at the same cost. So it is necessary to balance these to minimize the cost of the system.

Link budget is necessary in many radio communication systems other than satellite communication systems. A simple analysis is depicted in figure 6.1. Link budget calculations are also used for calculating the range of base station coverage. Link budget calculations are also important in wireless survey tools. These wireless survey tools will not only look at the way radio signals propagate, but also at the power levels, antennas and receiver sensitivity levels required to provide the required link quality [207].

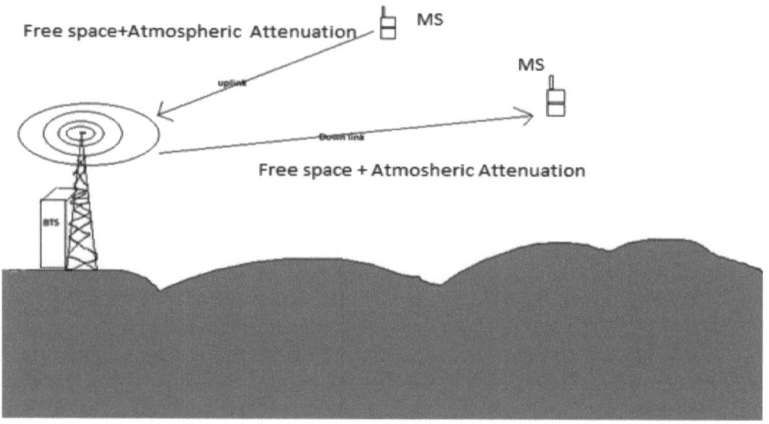

Figure 6.1 Illustration of Link Budget in Mobile Communication

The literature survey carried out during current research on link budget is mentioned in the following section.

6.2 COVERAGE AREA

A cellular network is a radio network circulated over small land areas called cells. A cell corresponds to the covering area of single BS transmitter or a small collection of many transmitters. These type of small cells are joined together to provide radio coverage over a large geographic area. This cellular radio network enables a large number of portable transceivers (e.g., mobile phones, pagers, etc.) to communicate with each other and with fixed transceivers and telephones anywhere in the network, via base stations. The cell and network coverage depend mainly on natural factors such as geographical propagation conditions, and on human factors such as the landscape (urban, suburban, rural), subscriber behaviour etc. The quality of the coverage in the mobile network is measured in terms of location probability. For that, the radio propagation conditions have to be predicted as accurately as possible for the region. Three main mechanisms that impact the signal propagation [186] are: Reflection, Diffraction, and Scattering. Reflection occurs when the electromagnetic wave strikes against a smooth surface, whose dimensions are large compared with the signal wavelength. Diffraction occurs when the electromagnetic wave strikes a surface whose dimensions are larger than the signal wavelength, new secondary waves are generated (shadowing). Scattering happens when a radio wave strikes against a rough surface whose dimensions are equal to or smaller than the signal wavelength.

The radio planners can either develop their own propagation models for different areas in a wireless network, or they can use the already existing propagation, which are non specific in nature and are used for a whole area. The advantage of developing a own model is that it will be more accurate, but time-consuming to construct. Usage of the existing propagation models is economical from the time and money perspective, but the accuracy of these models is suffer when they are used other than specific environments.

The empirical models uses existing equations obtained from results of several measurement efforts. Some of the path loss models are discussed in the current research. Those are Simplified Path Loss Model, Stanford University Interim (SUI) Model, Okumura's Model, Hata Model, COST231 Extension to Hata Model, ECC-33 model, Walfisch- Bertoni Model, Longley rice model, Egli Propagation Model, Bullington model, Epstein-Peterson model etc.

The above mentioned all the models are designed by calculating field data in different environments. Path loss determines the cell ranges. For GSM there are three cell ranges exist. Those are:

- Large cells, cell radius is 1 Km and normally it exceeds 3 Km.
- Small cells, cell radius is in between 1 Km and 3 Km.
- Microcells, of radius in the range of 200 m – 300 m.

The propagation in the above three cell sizes is determined by diffraction and scattering [145].

The link budget plays vital role in deciding the coverage area at network planning level. Link budget parameters has been analysed by several researchers for solving constraints of path loss during uplink and downlink. The radio system is designed in such a way that when the link budget for uplink and downlink put together, they should have the same maximum allowable path loss in both directions which is used to predict coverage area of BTS. To predict efficient coverage area three base stations has been selected for field data collection. The data has taken at Narnaul region. The methodology used for efficient prediction of coverage area is shown in fig. 6.2.

The pathloss readings have been collected by considering different climatic conditions. Those readings are compared with different predicted values by different path loss models (Discussed in chapter 4). By using this predicted path loss and measured path loss during test drive, a difference in link budget, i.e between uplink and downlink has been proposed. Addition of this extra loss in link budget, a new link budget has been proposed.

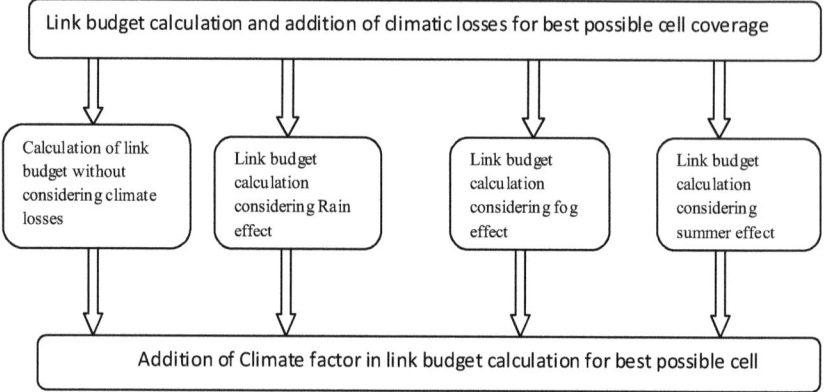

Figure 6.2 Methodology used for Prediction of Optimum Coverage Area

6.3 LINK BUDGET AND ITS CALCULATIONS

Link budget is calculation of all the gains and losses (total power calculations) in a transmission system. The link budget looks at the elements that will determine the signal strength arriving at the receiver. Link budget calculations are used for determining the power levels required for cellular communications systems and for investigating the base station coverage. In wireless communication systems, for uplink and downlink, the transmitting signal at a given frequency should provide coverage over a predetermined service area [240]. The link budget is based on the following parameters [79], [216]:

- Transmitter power.
- Antenna gains (both transmitter antenna gain and receiver antenna gain).
- Antenna feeder losses.
- Path loss
- Receiver sensitivity

The link budget general formula is given as:

$$ReceivedPower(dBm) = TransmittedPower + Gains - Losses \qquad (6.1)$$

In basic calculation of link budget equation it is supposed that the power spreads out equally in all directions from the source transmitter (isotropic antenna).

This is fine for theoretical calculations, But not appropriate for field calculations. A typical link budget equation for a radio communications system is given below:

$$P_{RX} = P_{TX} + G_{TX} + G_{RX} - L_{TX} - L_{FS} - L_{FM} - L_{RX} \qquad (6.2)$$

here,
P_{RX} = received power (dBm)
P_{TX} = transmitter output power (dBm)
G_{TX} = transmitter antenna gain (dBi)
G_{RX} = receiver antenna gain (dBi)
L_{TX} = transmit feeder and associated losses (feeder, connectors, etc.) (dB)
L_{FS} = free space loss or path loss (dB)
L_{FM} = many-sided signal propagation losses (these include fading margin, polarization mismatch, losses associated with medium through which signal is travelling and other losses) (dB)
L_{RX} = receiver feeder losses (feeder, connectors, etc.) (dB)

The objective of link budget calculation is to balance the uplink and down link. The uplink and downlink budget is shown in fig. 6.3. The coverage area prediction will use well known mathematical models and terrain data [219]. The power of BTS can be adjusted to balance the whole link. The power balance (uplink and down link) decide the cell range. In this work, two conditions are focused. These are:

Condition1: The down link is greater than the uplink [223]: It results in Range of BTS greater than Range of MS, Call connection will be dropped on uplink after initiation of handover, and Coverage area is smaller in reality than the prediction. This condition is most frequent.

Condition2: The uplink is greater than the down link: It results in Range of BTS less than Range of MS and No coverage problem from MS to BTS. The coverage is limited by the uplink because of the maximum available transmitting power of the mobile; the downlink sets limitations on the capacity due to the increasing interference [86].

From the above statement it is clear that the condition uplink > down link, is better than the condition uplink < down link.

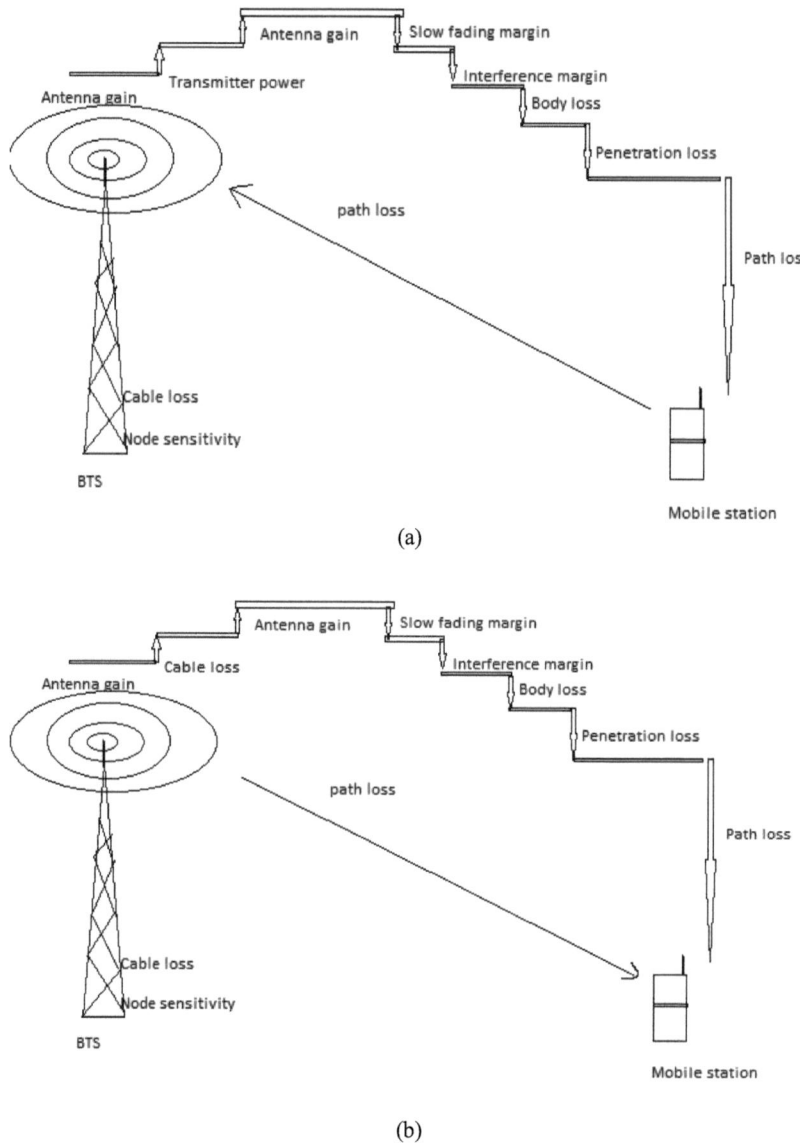

Figure 6.3 (a,b) Uplink & Downlink Budget

6.3.1 Important Parameters of Link Budget Calculations

The GSM link budget parameters are as follows [49], [79]:-

6.3.1.1 Receiver Sensitivity

The receiver sensitivity is one of the most important parameter to calculate link budget. It gives you the information about the minimum value of power that is needed to complete radio link. Low receiver sensitivity indicates good radio receiver. It can be expressed as:

Receiver Sensitivity (dBm) =
Thermal Noise (dBm/Hz) + (E_b/N_o)(dB) + information or Data Rate(dB-Hz) (6.3)

here,

E_b/N_o (i.e. output signal to noise ratio)

The required average E_b/N_o is a common measure of modem efficiency.

6.3.1.2 MS Sensitivity

Mobile station sensitivity (MS sensitivity) depends on the receiver noise figure and minimum level of E_b/N_o (output signal to noise ratio). MS Sensitivity can be calculated Using GSM specifications. The recommended values of MS sensitivity in GSM 900MHz and 1800MHz are –102 dBm and –100 dBm respectively.
The mathematical formula to calculate MS sensitivity is

$$MS_{Sensitivity} = 10\log_{10}(KTB) + (E_b/N_0) + NoiseFigure_{RX} \quad (6.4)$$

6.3.1.3 BTS Sensitivity

It also depends on the receiver noise figure and minimum level of E_b/N_o. The sensitivity of the base station is specified by GSM recommendations and is calculated in the same method as the MS sensitivity.

$$BTS_{Sensitivity} = 10\log_{10}(KTB) + (E_b/N_0) + Noise\ Figure_{RX} \quad (6.5)$$

6.3.1.4 MS & BTS Antenna Gain

The transmitter and receiver antenna gains are important calculations on a radio link. The antennas used for MS and BTS have considerably different gain levels. Transmitting antenna gain results from the focusing of emitted power in particular directions rather than from an increase in the emitted power [79].

6.3.1.5 Diversity Gains

Diversity gain is nothing but uplink Signal to noise ratio. Radio Frequency link balance depends on 3 factors which are BTS transmitter power, BTS combiner loss and BTS receiver diversity gain. SNR is generally lower on the uplink than the downlink because diversity is used in that direction but not the other. Since the diversity gain can depend on operating frequency, uplink SNR may differ for the two bands, while downlink SNR is the same for the two bands.

6.3.1.6 Feeder & Connector Loss

To complete the radio link in the transmitter side or in the receiver side it is necessary to use cables and connectors for proper power transmission. So it is essential to include these losses in power budget calculation. The Cable attenuation is usually extracted in loss (dB) per 100 m. So in the calculation the actual length of the cable should be multiplied by this value to get the theoretical loss taking place in the cable. Sometimes, the theoretical loss may go beyond the required value, so preamplifiers (also known as masthead amplifiers) may be used to counter the cable loss. Connector losses are very low comparison with the cable loss. [79]

6.3.1.7 Pre amplifier & Booster

To minimize cable loss pre amplifiers (Mast Head Amplifiers (MHA)) must be used at the receiver side of the link. MHAs improve the uplink link budget by compensating for Node B cable losses and improving the Node B receiver noise figure. The improvement in link budget may be studied using Friis' equation. In reality the cascaded receiver sub-system includes the MHA, feeder cable and a bias-T. The bias-T is used to provide power to the MHA by injecting a voltage into the feeder cable. This receiver sub-system is given in the figure below.

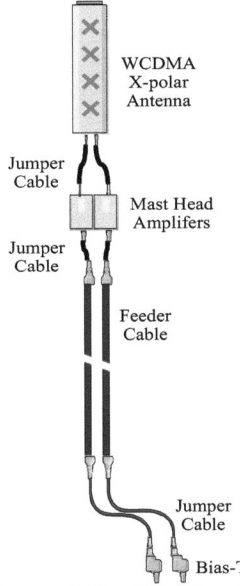

Figure 6.4 Mast Head Amplifier

6.3.1.8 Interference Degradation Margin

This is the difference between the received signal power and receiver threshold value. Different values are used for different types of regions, such as 2 dB for dense urban or 1 dB for urban.

6.3.1.9 Polarization Loss

Also important is the orientation of transmit and receive antennas relative to one-another. The electromagnetic waves which antennas radiate and receive have two components – an electric field and a magnetic field. These two fields travel through space at right angles (90°) to one another. The orientation of the electric field designates the "polarization" of the wave radiated by that antenna. Radio waves are polarized horizontally, vertically, or circularly. To maximize energy transfer from the transmitter to the receiver, the antenna polarizations need to match, regardless of which polarization is chosen. If it is not matched polarization losses occur in the communication system.

6.3.2 Uplink Budget and Coverage area

The figures 6.5 (a) and 6.5 (b) represent the process of uplink budget. In the calculation of uplink budget, first Tx EIRP has been calculated and with the help of EIRP the receiver sensitivity has been calculated. For calculation of the total allowable path loss, the total gains and losses has been taken.

6.3.2.1 Transmitting End

$$EIRP = P_{Tx} + L_{AF} + G_{ME} \qquad (6.6)$$

Table 6.1 Transmitter Side Specifications (Uplink)

Transmitter(Mobile equipment (ME) or MS)	Parameter Value
1. Transmitter power of ME (P_{Tx})	33dBm
2. MS or ME antenna gain (isotropic antenna) (G_{ME})	0
3. Connector loss or Antenna feeder loss (L_{AF})	0
4. Effective isotropic radiated power (EIRP= $P_{Tx} + L_{AF} + G_{ME}$)	33 dBm
5. Mobile station antenna height (h_m)	3 m

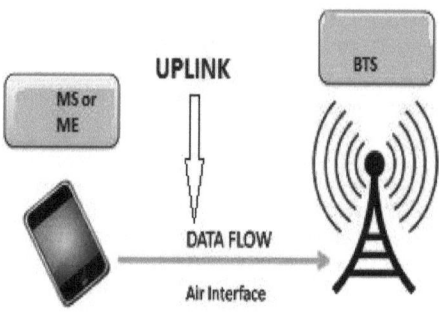

6.5 (a)

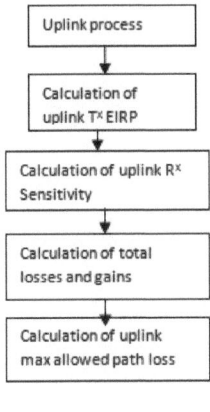

6.5 (b)

Figure 6.5 (a,b) Uplink Budget & Flow chart (Uplink budget)

6.3.2.1 Receiving End

$$R_s = EIRP - L_p - I_M - F_M - L_C + G_{BTS} \qquad (6.7)$$

Table 6.2 Receiver side specifications (uplink)

Receiver (BTS)	Parameter value
1. Receiver sensitivity (R_S)	-123.4dBm
2. Body loss (B_L)	3dB
3. BTS receiving antenna gain (G_{BTS})	18dBi
4. Interference margin (I_M)	2
5. Fast fade margin (F_M)	5dB
6. Connector loss (L_C)	4dB
7. Base station antenna height (h_b)	45m

Using equation (6.6)   EIRP= 33-0-0 = 33dB

Using equation (6.7) $R_s = EIRP - L_p - I_M - F_M - L_C + G_{BTS}$

-123.4 =33-L_p-2-5-4+18

Therefore L_P= 163.4 dB

6.3.3 Down Link Budget and coverage area

The figures 6.6 (a) and 6.6 (b) represent the process of downlink budget. In the calculation of downlink budget, after calculating the maximum allowable path the prediction of coverage area has been proposed. For calculation of the total allowable path loss, the total gains and losses have been taken.

6.3.3.1 Transmitting End

$$EIRP = P_{TXB} - L_{cableB} - L_c + G_{TXB} \qquad (6.8)$$

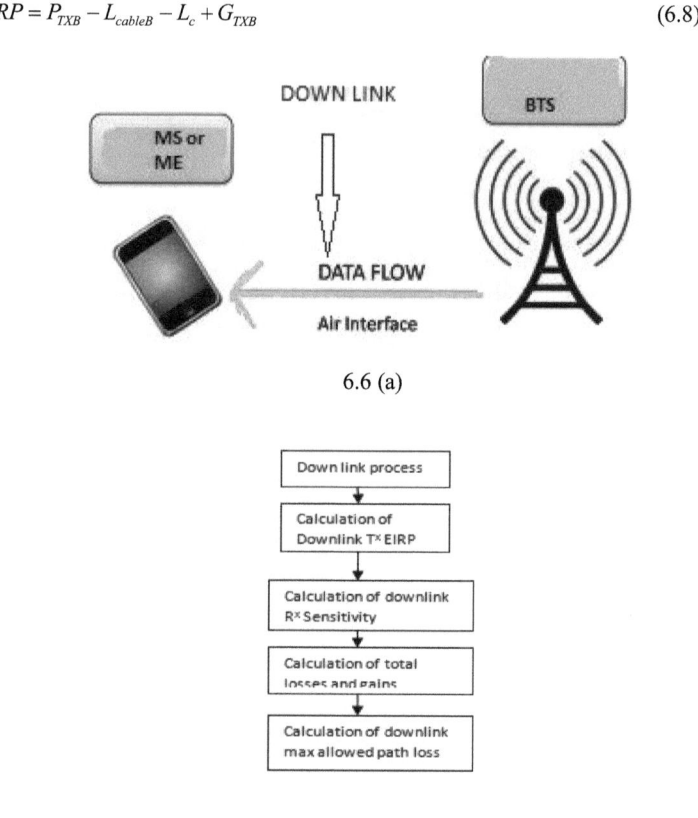

Figure 6.6 (a,b) Downlink Budget & Flow Chart (Downlink Budget)

Table 6.3 Transmitter Side Specifications (Down link)

Transmitter(BTS)	Parameter Value
1. Out put power of BTS (P_{TxB})	46 dBm (general value)
2. Transmitter antenna gain (G_{TXB})	18 dB
3. Cable loss (L_{CableB})	2 dB
4. Combiner loss(L_c)	2 dB
5. $EIRP = P_{TXB} - L_{cableB} - L_c + G_{TXB}$	60 dB

6.3.3.2 Receiver End

Table 6.4 Receiver side specifications (Down link)

Receiver (MS or ME)	Parameter value
1. Mobile station Sensitivity (R_{SM})	-107.5 dBm
2. Body loss (B_{LM})	0 dB
3. MS receiving antenna gain (G_{BTS})	0 dB
3. Interference margin (I_M)	3 dB
4. Control channel overhead (L_{Coh})	1 dB

$$R_s = EIRP - L_p - B_{LM} - I_m - L_{Coh} + G_{MS} \qquad (6.9)$$

R_S= -107.5= 60-Lp-3-1-0-0

L_p= 163.5

It has been already discussed in chapter 4 that Okumura propagation path loss model is suitable propagation path loss model for the Narnaul, Haryana. The path loss equation from equation 4.7 is:

PL Okumura (dB) = $L_F + A_{mu}(f,d) - G(h_{te}) - G(h_{re}) - G_{area}$

From above calculations (Link budget analysis) the path loss value is 163.5dB.

At 1800MHz

BTS height (h_b) = 45m

Lf= $20*(\log_{10}(d))+97.5555$;

PL Okumura=$20*(\log_{10}(d))+144.5118$;

By substituting PL Okumura=163.5

$\log_{10}d=(163.5-144.5118)/20$

The coverage value is d=8.904 km

At 1800MHz

BTS height (h_b) = 35m

Lf= $20*(\log_{10}(d))+97.5555$;

PLOkumara= $20*(\log_{10}(d))+146.6947$;

By substituting PLokumura=163.5

$\log_{10}d=(163.5-146.6947)/20$

The coverage value is d=6.9225 km

At 900MHz

BTS height (h_b) = 45m

Lf= $20*(\log_{10}(d))+91.5349$

PL Okumura=$20*(\log_{10}(d))+138.4912$

By substituting PL Okumura = 163.5

$\log_{10}d=(163.5-138.4912)/20$

The coverage value is d=17.900 km

At 900MHz

BTS height (h_b)=35m

Lf= $20*(\log_{10}(d))+ 91.5349$;

PL Okumara= $20*(\log_{10}(d))+ 140.6741$

By substituting PL Okumura=163.5

$\log_{10}d=(163.5-140.6741)/20$

The coverage value is d=13.84 km

6.4 EFFECT OF CLIMATIC CONDITIONS ON LINK BUDGET

The link budget calculations are mainly depends on the path loss. And the path loss depends on climatic conditions and nature of the area. It has been clearly discussed in chapter 5. In the same vein calculation of Link budget depends on path loss. On the basis of climatic conditions in the same chapter Developed Okumura model is given to calculate Path loss value. The path loss equation from equation 5.6 is:

Developed Okumura model= Okumura model + AF + CF

Where AF is area factor it can vary according to area and it is equal to 6.5137 in mentioned environment (Narnaul, Haryana Region) CF is climate factor:

$$CF = \begin{cases} 0 & For & ORDINARY\ CLIMATE \\ \dfrac{10\log_{10}(\varepsilon)}{V}d & For\ densefog & <50m & VISIBILITY \\ Sa_R * d & Heavy rain & \geq 100mm & RAINFALL \end{cases}$$

6.4.1 Calculation of Link Budget & Coverage Area in Summer and winter

The effect is already discussed in previous chapters. On the basis of those discussions the following calculations have been carried out

For normal climatic conditions CF=0;

AF=6.5137

Now substituting the path loss = 163.5dB after the link budget analysis:

At 1800MHz

BTS height (h_b) = 45m

Lf= $20*(\log_{10}(d))+97.5555$;

Developed Okumura=$20*(\log_{10}(d))+144.5118+6.5137$;

By substituting Developed Okumura=163.5

$\log_{10} d = (163.5-144.5118-6.5137)/20$

The coverage value is d=4.2046 km

At 1800MHz

BTS height (h_b) = 35m

Lf= $20*(\log_{10}(d))+97.5555$;

Developed Okumura= $20*(\log_{10}(d))+146.6947+6.5137$

By substituting Developed Okumura=163.5

$\log_{10} d = (163.5-146.6947-6.5137)/20$

The coverage value is d=3.2702 km

At 900MHz

BTS height(h_b)=45m

Lf= $20*(\log_{10}(d))+ 91.5349$

Developed Okumura=$20*(\log_{10}(d))+ 138.4912163.5+6.5137$

By substituting Developed Okumura=163.5

$\log_{10} d = (163.5-138.4912-6.5137)/20$

The coverage value is d=8.4092 km

At 900MHz

BTS height (h_b) =35m

Lf= 20*(log$_{10}$(d))+ 91.5349;

Developed Okumura= 20*(log$_{10}$(d))+ 140.6741+6.5137

By substituting Developed Okumura=163.5

Log$_{10}$d=(163.5-140.6741-6.5137)/20

The coverage value is d=6.5405 km

Table 6.5 Coverage Area Calculations in Summer & Winter

Frequency BTS Antenna height (h_b)	1800 MHz	900 MHz
45m	4.2046 km	8.4092 km
35m	3.2702 km	6.5405 km

6.4.2 Calculation of Link Budget and Coverage Area in Heavy Fog (visibility=30m)

For heavy fog climatic conditions CF=2.2185

AF=6.5137

Now substituting the path loss= 163.5dB after the link budget analysis:

At 1800MHz

BTS height (h_b) = 45m

Lf= 20*(log$_{10}$(d))+97.5555;

Developed Okumura=20*(log$_{10}$(d))+144.5118+6.5137 +2.2185 ;

By substituting Developed Okumura=163.5

Log$_{10}$d=(163.5-144.5118-6.5137-2.2185)/20

The coverage value is d = 3.2569km

At 1800MHz

BTS height(h_b)=35m

$L_f = 20*(\log_{10}(d)) + 97.5555$;

Developed Okumura= $20*(\log_{10}(d)) + 146.6947 + 6.5137 + 2.2185$

By substituting Developed Okumura=163.5

$\log_{10}d = (163.5 - 146.6947 - 6.5137 - 2.2185)/20$

The coverage value is d=2.5331km

At 900MHz

BTS height (h_b) =45m

$L_f = 20*(\log_{10}(d)) + 91.5349$

Developed Okumara= $20*(\log_{10}(d)) + 138.4912163.5 + 6.5137 + 2.2185$

By substituting Developed okumura=163.5

$\log_{10}d = (163.5 - 138.4912 - 6.5137 - 2.2185)/20$

The coverage value is d=6.5137km

At 900MHz

BTS height (h_b) = 35m

$L_f = 20*(\log_{10}(d)) + 91.5349$;

Developed Okumara= $20*(\log_{10}(d)) + 140.6741 + 6.5137 + 2.2185$

By substituting Developed okumura=163.5

$\log_{10}d = (163.5 - 140.6741 - 6.5137 - 2.2185)/20$

The coverage value is d=5.0662 km

Table 6.6 Coverage Area in Foggy Days

Frequency BTS Antenna height (h_b)	1800 MHz	900 MHz
45m	3.2569 km	6.5137km
35m	2.533km	5.662km

6.4.3 Calculation of Link Budget and Coverage Area in Heavy Rain (100mm/hr)

For heavy rain conditions CF=0.0068*d

AF= 6.5137

Now substituting the path loss= 163.5dB after the link budget analysis:

At 1800MHz

BTS height(h_b)=45m

Lf= 20*($\log_{10}(d)$)+97.5555;

Developed Okumura=20*($\log_{10}(d)$)+144.5118+6.5137 +0.068*d ;

Developed Okumura=20*($\log_{10}(d)$)+ 151.0255+0.068*d ;

By substituting Developed Okumura=163.5

$20\log_{10}d$+0.068d=(12.4745)

The coverage value is d=4.1908km

At 1800MHz

BTS antenna height(h_b) = 35m

Lf= 20*($\log_{10}(d)$)+97.5555;

Developed Okumura= 20*($\log_{10}(d)$)+146.6947+6.5137 +0.068*d

Developed Okumura= 20*($\log_{10}(d)$)+ 153.2084 +0.068*d

By substituting Developed Okumura=163.5

$20\log_{10}d + 0.068d = (10.2916)$

The coverage value is d=3.2619km

At 900MHz   **The rain attenuation is negligible [39]**

i.e CF≈0

BTS antenna height (h_b)=45m

Lf= $20*(\log_{10}(d)) + 91.5349$

Developed Okumara=$20*(\log_{10}(d)) + 138.4912163.5+6.5137$

By substituting Developed Okumura=163.5

$\log_{10}d = (163.5-138.4912-6.5137)/20$

The coverage value is d=8.4092 km

At 900MHz

BTS antenna height (h_b) = 35m

Lf= $20*(\log_{10}(d)) + 91.5349$;

Developed Okumura= $20*(\log_{10}(d)) + 140.6741+6.5137$

By substituting Developed Okumura=163.5

$\log_{10}d = (163.5-140.6741-6.5137)/20$

The coverage value is d=6.5405 km

Table 6.7 Coverage Area Calculation in Rainy Days

Frequency BTS Antenna height (h_b)	1800 MHz	900 MHz
45m	4.19km	8.4092km
35m	3.26km	6.5405 km

6.4.4 Calculation of Link Budget and Coverage Area Including all Climatic Effects in Narnaul (Haryana, India)

Total path loss = Developed Okumura model including all losses
= Okumura model + AF + Rain attenuation + Fog attenuation

For heavy rain CF=0.0068*d

For heavy fog climatic conditions CF=2.2185 and

AF= 6.5137

Now substituting the path loss= 163.5dB after the link budget analysis:

At 1800MHz

BTS height (h_b) =45m

Lf= $20*(\log_{10}(d))$+97.5555;

Developed Okumura=$20*(\log_{10}(d))$+144.5118+6.5137 +2.2185+0.068*d ;

163.5=$20*(\log_{10}(d))$+ 153.2440+0.068*d ;

10.2560=$20*(\log_{10}(d))$ +0.068*d ;

D=3.2486 km

At 1800MHz

BTS antenna height (h_b) = 35m

Lf= $20*(\log_{10}(d))$+97.5555;

Developed Okumura= $20*(\log_{10}(d))$+146.6947+6.5137+2.2185 +0.068*d

163.5= $20*(\log_{10}(d))$+ 155.4269+0.068*d

By substituting Developed Okumura=163.5

$20\log_{10}d$+0.068d= (10.2916)

The coverage value is d=2.5281km

The rain attenuation is negligible i.e. CF due to rain≈0

At 900MHz

BTS height (h_b)=45m

Lf= $20*(\log_{10}(d))+ 91.5349$

Developed Okumura=$20*(\log_{10}(d))+ 138.4912163.5+6.5137+2.2185$

By substituting Developed Okumura=163.5

$\log_{10} d= (163.5-138.4912-6.5137-2.2185)/20$

The coverage value is d=6.5137km

At 900MHz

BTS height (h_b) = 35m

Lf= $20*(\log_{10}(d))+ 91.5349$;

Developed Okumura= $20*(\log_{10}(d))+ 140.6741+6.5137+2.2185$

By substituting Developed Okumura=163.5

$\log_{10} d=(163.5-140.6741-6.5137-2.2185)/20$

The coverage value is d=5.0662 km

Table 6.8 Coverage Area Calculations by Using Developed Okumura Model

Frequency BTS Antenna height (h_b)	1800 MHz	900 MHz
45m	3.241 km	6.5137km
35m	2.5281km	5.0662km

6.5 CONCLUSION

Drive test is performed at each selected site, during all seasons (summer, winter, heavy rain and heavy fog) which give maximum attenuation practised by mobile station in a radio cell. Based on the field data, it has been observed that difference in path loss without considering climatic losses and path loss during those effects indicates that there is a need of introducing a correction factor in link budget. The amount of climate factor calculated has been used as correction factor and it is added in the link budget. The maximum allowable path loss with and without Climate factor has been analysed for proper link budget design. From the analysis, it has been observed that there is an inverse relationship between path loss and coverage area i.e. increase in path loss results decrease in coverage area as shown in tables from 6.5 to 6.8. In order to balance the loss due to climate and environment there is a need to increase the link budget.

CHAPTER 7

CONCLUSION AND FUTURE WORK

This chapter gives brief summary of the results obtained in the earlier chapters as well as their importance. In present chapter conclusion and recommendations for future work have been presented. The radio propagation is necessary for wireless communication and it mainly depends on type of environment, speed of mobile terminal and frequency of transmission. Mathematical modelling along with accurate characterization of radio channel is essential to predict signal coverage and path loss during transmission. In the current research work, it has been investigated that spatial particle density increases according to the climate changes. The increment of particle dust density is mainly due to atmosphere. Specific attenuation of particle increases with the size of the dust particles and totally depends on the complex permittivity of the dust particle. In this study it has been observed that Atmospheric losses are caused by scattering because the wavelengths of GSM frequencies are larger than particles size parameter. It is found that particle density in the atmosphere has effect on signal strength.

7.1 RESULTS & CONCLUSION

The results of presently carried research work have relevant application in designing field propagation model for the surrounding of Narnaul (Haryana, India) under different environmental conditions. Many field propagation models such as free space path loss, Hata path loss, Okumura, Egli, ECC-33, Cost-231, LEE model, W-I, SUI and Bertoni at GSM frequency band has been analysed. In present research work the comparative analysis between path loss models and field measured data carried out at Narnaul (Haryana, India). The analysis of every field propagation model depends on environmental conditions under which the field data collection is carried. The field propagation model gives accurate results if it is used under similar conditions in which the model was designed; if not the results may vary. In the current work five different base stations operating at GSM frequency band has been considered for field data measurements to develop a propagation model for Narnaul region (Haryana, India). Out of those five cell sites three are from urban area and two are from suburban and rural area.

Table 7.1 Difference between Measured Field Data to Path Loss Model

Path loss Model	Difference between measured Field data to Path loss Model
Free Space Path loss model	40dB
W-I Path loss Model	34dB
Lee Path loss Model	20dB
Egli Path loss Model	50dB
Bertoni Path loss Model	30dB
Okumura Path loss Model	6dB
Cost-231 Path loss Model	20dB
ECC-33 Path loss Model	15dB
SUI Path loss Model	28dB
Hata Path loss Model	20dB

The field data collected during test drive in three cell ids in urban area has been compared with different path loss prediction models. After observing the various field propagation models, it has been observed that no propagation model predict precisely for propagation terrain of Narnaul (Haryana). This is due to the fact that propagation of radio waves is strongly influenced by the nature of terrain particularly the size and density of buildings and climatic conditions. The approximated error between measured field data and pathloss model is shown in Table 7.1. The Okumura path loss model yields better results than other propagation models in the terrain of Narnaul (South Haryana) which is also clear from Table 7.1.

The model indicates a good agreement with the measured path loss. It is due to the fact that Okumura model takes all the parameters into the account such as free space loss, median attenuation relative to free space, transmitting/receiving antenna gain and gain due to environment. Some modification is still needed in the Okumura model to get better results in the mentioned environment.

Table 7.2 Average Error between currently measured data with Okumura and Developed Okumura Model

Model	Average Error (dB) summer	Average Error (dB) Winter	Average Error (dB) Heavy rain	Average Error (dB) Heavy fog
Okumura	4.26	4.16	5.26	5.71
Developed Okumura	0.253	0.35	0.73	0.13

To give modification in the Okumura model, drive test is performed during all seasons (summer, winter, heavy rain and heavy fog) at each selected site, which yields maximum attenuation generally, takes place in mobile station in a radio cell. The field measured data in different climatic conditions has been compared with the Okumura path loss model. On the basis of this analysis a pathloss prediction model has been proposed. The developed Okumura pathloss model contains area factor and climate factor along with the original Okumura model. With the help of developed Okumura model, the pathloss prediction is more precise as compared to the original Okumura model. From Table 7.2, it is observed that the average error between currently measured data with Okumura is more and that the average error between currently measured data with Developed Okumura Model is less.

Validation of developed Okumura model has been taken using the reference attenuation models for rain and fog. The reference rain attenuation equation is taken from M. Sridhar research paper and the reference fog attenuation is taken from Altshuler. In this investigation the difference in both attenuation equations is very small. Further the developed Okumura model is compared with field measured data taken at Hisar (Haryana, India). This investigation shows that the developed model is valid through out the Haryana and the area which has same climatic conditions similar to Narnaul.

Table 7.3 Coverage Area Calculations taking Different Parameters

Frequency & BTS height / Climate Condition	F=1800 MHz (h_b)=45m	F=1800MHz (h_b)=35m	F=900 MHz (h_b)=45m	F=900 MHz (h_b)=35m
Summer	4.2046 km	3.2702km	8.4092km	6.5405 km
Winter	4.2046 km	3.2702km	8.4092km	6.5405 km
Heavy fog winter	3.2569	2.533	6.5137	5.0662
Heavy rain	4.19	3.26	8.4092	6.5405

Drive test is performed at each selected site, during summer, winter, heavy rain and heavy fog conditions. Based on the field data, it has been observed that difference in path loss without considering climatic losses and path loss by considering those losses indicates that there is a need of introducing a correction factor in link budget. The amount of climate factor calculated has been used as correction factor and it is added in the link budget. The maximum allowable path loss with and without Climate factor has been analyzed for proper link budget design. The coverage area calculations at different parameters are given in Table 7.3 In order to balance the loss due to climate and environment there is a need to increase the link budget. By the analysis, it has been observed that there is an inverse relationship between path loss and coverage area i.e. increase in path loss results decrease in coverage area.

7.2. FUTURE WORK

In present analysis, climatic effects on radio wave propagation has been investigated and analysed. The effect on the variation of field strength of the signal is included as a main aim. The analysis on "Rate of fading" can deeply examined for further analysis of this phenomenon and would be advisable. In future, fuzzy logic approach can be implemented to determine the unknown environment attenuation. This would however require sophisticated equipment. The embedded system may be designed which can accommodate variable gain depending on tuning. Some other environmental factors that affect signal propagation should further be investigated.

Future work includes investigations about the influence of radio propagation modelling in classification of mutually exclusive strategies, such as different routing approaches, which are concerned in improving a given performance parameter.

Another area for investigation of local conditions on radio wave propagation could be comparative studies of climate effects on different kinds of modulation. The paramount aspect of the examination of physical phenomenon for instance climate effects, with respect to its influence on radio reception should be conducted on long term basis. The two different methods for measuring the field strength can be used, one is by communication receiver and another could be based on the use of an electric field strength measuring device specifically designed for this kind of task. It would be advisable for a radio planner at the Narnaul region in state Haryana to reduce the distance between BTS's to maintain the QoS during severe climate changes.

REFERENCES

1. Abdullahi, T.S., 2008. Network Optimization in a GSM System: A Case Study of MTN Nigeria Communications Limited. MSc Thesis, Electrical Engineering Department, Ahmadu Bello University Zaria.
2. Abhayawardhana V. S., I.J. Wassel, D. Crosby, M.P. Sellers and M.G. Brown, 2005. Comparison of empirical propagation path loss models for fixed wireless access systems. *61th IEEE Technology Conference*, pp. 73-77.
3. Adebayo T. L. and F.O. Edeko, 2006. Characterization of Propagation Path Loss at 1.8GHz:A Case study of Benin city, Nigeria. *Research Journal of Applied Sciences: Medwell online*.
4. Adegoke A. S., Babalola, and W.A.Balogun,2008. Performance Evaluation of GSM Mobile System in Nigeria. *The Pacific Journal of Science and Technology*, vol 9. no 2. pp. 436-442.
5. Agrawal D. P. and Q. Zeng, 2003. *Introduction to Wireless and Mobile Systems*. Thomson Books / Cole, Chapter 3.
6. Ahmed H.Zahram, Ben Liang and Aladdin Dalch, 2006. Signal threshold adaptation for vertical handoff on heterogeneous wireless networks. *Mobile Networks and application*, vol.11, no.4, pp. 625- 640.
7. Ahmed M. I., Z. Nadir, 2010. Characterization of Pathloss using Okumura-Hata Model and missing Data Prediction for Oman. *IAENG Transactions on Engineering and Technology*, vol. (5), pp. 509-518.
8. Ahuja Kiran, Kumar Manoj, 2011. Significance of Empirical and Physical Propagation Models to Calculate the Excess Path Loss. Journal of Engineering Research and Studies, Vol.II, Issue II, pp. 1-6.
9. Ajay Mishra, 2007. *Advance Cellular Network Planning and Optimisation*, John Wiley and Sons, ISBN-10 0-470-01471-7 (HB),pp.197-303.
10. Alagu S., and Meyyappan T., 2011. Analysis of Handoff Schemes in Wireless Mobile Network. IJCES International Journal of Computer Engineering Science , Volume1 Issue2, pp. 1-95.
11. Alim M. A., Rahman M. M., Hossain M. M. Hossain, A. Al-Nahid, 2010. Analysis of Large-Scale Propagation Models for Mobile Communications in Urban Area. *International Journal of Computer Science and Information Security*, Vol. 7, No. 1, pp. 135-139.
12. Aliye "Ozge Kaya., Larry J. Greenstein., 2012. A New Path Loss Modeling Approach for In-Building Wireless Networks. *Globecom 2012 - Wireless Communications Symposium*. pp. 5255-5259.
13. Allnutt Jeremy, Dissanayake Asoka, Haidara Fatim, 2002 "Prediction Model that Combines Rain Attenuation and Other Propagation Impairments Along Earth- Satellite Paths. NASA's Advanced Communications Technology Satellite (ACTS), Issue 2, Fall 2002.
14. Almes P. et.al, 2007. Survey of Channel and Radio Propagation Model for wireless MIMO systems. *Eurosip Journal of Wireless Communication Technology*, vol. 17, pp. 1-19.
15. Alotaibi F. D., A. Abdennour and A. A. Ali, 2009. A real-time intelligent wireless mobile station location estimator with application to TETRA network.

IEEE Transactions on Mobile Computing, vol. 8, pp.1495-1509.
16. Alotaibi F. D., and A. A. Ali, 2008. Tuning of Lee path loss model based on recent RF measurements in 400 MHz conducted in Riyadh city, Saudi Arabia. *The Arabian Journal for Science and Engineering,* vol 33, no 1B, pp. 145–152.
17. Alotaibi F. D., and Adel A. Ali, 2006. TETRA Outdoor Large- Scale Received Signal Prediction Model in Riyadh City, Saudi Arabia. *Proceeding of IEEE WAMICON'06,* Clearwater, Florida, USA.
18. Andreas F. Molisch, Tufvesson F., Karedal J. and Mecklen C, 2009. A survey on vehicle-to-vehicle propagation channels. *IEEE Wireless Communication*, vol. 16, no. 6, pp. 12–22.
19. Armoogum V., Soyjaudah K.M.S., Fogarty T. and Mohamudally N. , 2007. Comparative Study of Path Loss using Existing Models for Digital Television Broadcasting for Summer Season in the North of Mauritius. *Proceedings of Third Advanced IEEE International Conference on Telecommunication-Mauritius* ,vol. 4, pp. 34-38.
20. Arne Schmitz and Martin Wenig , 2008. The Effect of the Radio Wave Propagation Model in Mobile Ad Hoc Networks. *Proceedings Of World Academy Of Science, Engineering And Technology,* vol. 36 , pp.28-31.
21. Arunad J.F. and Port R. E., 1985. A comparison of prediction models for 800 MHz mobile radio propagation. *IEEE Trans.*, vol. 34(4), pp.149-153.
22. Asoka Dissanayake1, Jeremy Allnutt2, Fatim Haidara "Prediction Model that Combines Rain Attenuation and Other Propagation Impairments Along Earth-Satellite Paths" Online Journal of Space Communication 1546 – 1558 Issue Date :Oct 1997, Antennas and Propagation, IEEE Transactions on (Volume:45 , Issue: 10)
23. Athanasiadou G. E., A. R. Nix, and J. P. Mc Geehan, 2000. A microcellular ray-tracing propagation model and evaluation of its narrowband and wideband predictions. *IEEE Journal on Selected Areas in Communications, Wireless Communications series,* vol. 18, pp. 322–335.
24. Ayyappan K., Dananjayan P., Propagation Model for Highway in Mobile Communication System. Ubiquitous Computing and Communication Journal, Vol. 3, No. 4, pp. 61-66.
25. Bach Andersen J., 1994. Issues and challenges of propagation studies for mobile networks. *Proceeding of Personal, Indoor and Mobile Radio Conference PIMRC'94,* pp. 1285-1291.
26. Bai F., L. Cheng, B. E. Henty, and D. D. Stancil, 2008. Highway and rural propagation channel modeling for vehicle-to-vehicle communications at 5.9 GHz. *Proceeding of IEEE Antennas Propagation* , pp. 1–4.
27. Ball C. F., E. Humburg, K. Ivanov, and F. Treml, 2006. Comparison of IEEE 802.16 WiMAX scenarios with fixed and mobile subscribers in tight reuse. *European Transactions on Telecommunications,* vol. 17, no. 2, pp. 203–218.
28. Basharat A. , I.A. Khokhar and S. Murtaza ,2008. CDMA versus IDMA for subscriber cell density. *International Conference on Innovations in Information Technology*, vol. 12, pp. 520 – 524.
29. Batariere, M.D et al., 2004. Seasonal Variations in Path Loss in the 3.7 GHz Band. IEEE Radio and Wireless Conference, pp. 399 – 402.
30. H. Bertoni, W. Honcharenko, L. R. Maciel and H. Xia, "*UHF propagation prediction for wireless personal communications*", *Proc. IEEE*, vol. 82, pp. 1333

-1359, Sept. 1994.
31. Bilal Haider, M. Zafrullah and M. K. Islam., 2009. Radio Frequency Optimization & QoS Evaluation in Operational GSM Network. *Proceedings of the World Congress on Engineering and Computer Science*. vol. 1, ISBN:978-988-17012-6-8.
32. Black Peter J. and Qiang Wu. Link Budget of CDMA2000 1xEV- Wireless Internet Access System Qualcomm Incorporated, 5775 Morehouse Drive, San Diego, CA 92121.
33. Blake, Roy, "Wireless Communication Technology" Delmar Thomas Learning, 2001.
34. Bruce L.C., 2006. Prediction of Seasonal Trends in Cellular Dropped Call Probability. IEEE Proceedings of the International Conference on Electro/Information Technology, pp. 613-618.
35. Bryant G. H., I. Adimula, C. Riva and G. Brussaard, Rain attenuation statistics from rain cell diameters and heights (pages 263–283) Article first published online: 4 MAY 2001 | DOI: 10.1002/sat.673
36. Cakaj Shkelzen, 2009. Rain Attenuation Impact on Performance of Satellite Ground Stations for Low Earth Orbiting (LEO) Satellites in Europe. Int. J. Communications, Network and System Sciences, 2009, pp. 480-485.
37. Carstensen L. W., C.W. Bostian, and G.E. Morgan, 2001. Combining electromagnetic propagation, geographic information systems, and financial modeling in a software package for broadband wireless wide area network design. *Proc. ICEAA01*, pp. 799-810.
38. Casaravilla J., Dutra G., Pignataro N. and Acuna J., 2009. Propagation Model for Small Macro cells in Urban Areas. *IEEE transactions on vehicular technology*, vol. 58, no. 7, pp. 20-25.
39. T. S. Chu "Rain-induced cross polarization at centimeter and millimeter wavelengths", *Bell System Tech. J.*, vol. 53, pp.1557-1578 1974.
40. Chen, H., Dai, J. & Liu, Y. Effect of fog and clouds on the image quality in millimetre communications. International Journal of Infrared and Millimeter Waves, Vol.25, No.5, pp. 749-757, ISSN 1572-9559, 2004.
41. L. Chen, T. Sun, B. Chen, V. Rajendran, and M. Gerla, "A smart decision model for vertical handoff." The 4th Int'l Workshop on Wireless Internet and Reconfigurability (ANWIRE'04), May 2004..
42. Chew L, "Maps: Coordinate syatems," GPS workshop manual, 1998.
43. Christian Mätzler, Rudolf Hüppi, Review of signature studies for microwave remote sensing of snowpacks. Advances in Space Research Volume 9, Issue 1, 1989, Pages 253–265.
44. Christian M. Ho, et. al. " Estimation of Microwave Power Margin Losses Due to Earth's Atmosphere and Weather in the Frequency Range of 3–30 GHz" a report by Jet Propulation Laboratory, California Institute of Technology, Jan. 2004.
45. Chu T.S. and Greenstein Larry J., 1999. A Quantification of Link Budget Differences Between the Cellular and PCS Bands. IEEE transactions on vehicular technology, Vol. 48, No. 1, pp. 60-65.
46. Constantino Perez-Vegay., Jose Luis Garc´ıa Gy., and Jos´e Miguel L´opez Higueraz., 1997.A simple and efficient model for indoor path-loss prediction. *Meas. Sci. Technology*. vol.8, pp. 1166–1173.
47. Corazza G.E. & Others, "Characterization Of Handover Initialization in Cellular

Mobile Radio Networks", IEEE VTC 94, pp. 1869 –1872, 1994.
48. COST Action 231, 1999. Digital mobile radio towards future generation systems- final report. *Technical Report European Communities*, EUR 18957.
49. Crane R.K. and Blood D.W., 1979. Handbook for the Estimation of Microwave Propagation Effects - Link Calculations for Earth- Space Paths. Environmental Research and Technology Rpt. No.1, DoC. No. P-7376-TRL.
50. Crane R. K., 1980. Prediction of attenuation by rain. *IEEE Transactions on Communications,* vol.28, pp. 1727–1732.
51. Damosso (E.. Dependence of specific rain-attenuation and phase shift on electrical, meteorological and geometrical parameters.*CSEL T* Rapp. Tecnici, It. (sep. 1978),6, no 3, pp. 199–205
52. Dajab, D.D.. Characterisation and Modelling of 900MHz Indoor Wireless communication channels in the savannah region. PhD Dissertation, Electrical Eng. Dept. Ahmadu Bello University, Zaria. 45-52; 144-152, 2005.
53. Damosso, E., "Action COST 231: a commitment to the transition from GSM to UMTS," IEEE International Conference on Personal Wireless Communications, 18-19 Aug., 1994.
54. Dey Subhrakanti. and Evans J., 2007. Outage Capacity and Optimal Power Allocation for Multiple Time Scale Parallel Fading. *IEEE Transaction on Wireless Communication Systems,* vol. 6, no.7,pp. 2369-2373.
55. Desile G. Y., Lefevre J. P., Lecours M. and Chourinard J. Y.,1985. Propagation loss prediction: a comparative study with application to the mobile radio channel. *IEEE Transaction*, vol 32, no.2,pp. 86-95.
56. Desposito L, 2000. High-Speed Networks and 3G Wireless Hike Demands on Test Gear," Elect. Design, pp. 88.
57. De Bruyne J., W. Joseph, L. Martens, C. Olivier, and W. De Ketelaere, 2009. Field measurement and performance analysis of an 802.16 system in a suburban environment. *IEEE Transactions on Wireless Communications,* vol. 8, no. 3, pp. 1424–1434.
58. Dougherty H.T., and E.J. Dutton, 1978. Estimating Year-to-Year Variability of Rainfall for Microwave applications. IEEE Trans. Communication Vol. 26, No. 8 PP. 1321-1324.
59. Douglas O.R., 1973. Some Propagation Experiments Relating Foliage Loss and Diffraction Loss at X-Band and UHF Frequencies. IEEE Transactions on Vehicular Technology,Vol. 22 , Issue 4, pp. 114.
60. Elabdin Z., Md. Rafiqul Islam, Othman O. Khalifa, Hany Essam A Raouf, Momoh Jimoh E Salami., 2008. Development of Mathematical Model for the Prediction of Microwave Signal Attenuation due to Duststorm. *Proceedings of the International Conference on Computer and Communication Engineering* 978-1-4244-1692-9/08/$25.00 ©2008 IEEE, pp.1156-1161.
61. Elabdin Z., M. R. Islam., O. O. Khalifa., and H. E. A. Raouf., 2009. Mathematical Model for the Prediction of Microwave Signal Attenuation due to Duststorm. *Progress In Electromagnetics Research,* vol. 6, pp.139–153.
62. Electronic Communication Committee (ECC) within the European Conference of Postal and Telecommunications Administration (CEPT), 2003. The analysis of the coexistence of FWA cells in the 3.4 - 3.8 GHz band. *technical report, ECC Report 33.*
63. European Cooperation in the Field of Scientific and Technical Research EURO-

COST 231, 1991. Urban Transmission Loss Models for Mobile Radio in the 900 and 1800 MHz Bands. *Revision 2. The Hague*, September.
64. Eyo, O.E., Menkiti.A.I., Udo.S.O., 2003. Microwave Signal Attenuation in Harmattan Weather Along Calabar-Akampkpa Line-Of-Sight Link. *Turk Journal of Physics,* Vol.27, pp.153-160.
65. Fedi F, *Radio Science*, Radio Science, Volume 16, Issue 5, pages 731–743, September-October 1981.
66. Fedi (F.). Attenuation due to rain on a terrestrial path.*Alta Frequenza*, Ital. (1979),40, n° 4, pp. 47E-51E.
67. Folaponmile A and Sani M.S., July 2011. Empirical model for the rediction of mobile radio cellular signal attenuation in harmattan weather. *Information Technology Research Journal* , Vol 1(1) pp. 13 - 20.
68. Fraizer E.W., 1984. Handbook of Radio Wave Propagation Loss (100 – 10,000 MHz). A Technical Report, National Telecommunications and Information Administration, Annapolis Maryland.
69. Freeman, R. L. (2007). *Radio System Design for Telecommunications*, Third Edition, John Wiley&Sons, ISBN: 978-0-471-75713-9, New York
70. Fujimoto K. and Fujimoto Kuohei, 2001. Mobile Antenna System Handbook. Artech House Antenna and Propagation Library.
71. Galati, G., Dalmasso, I., Pavan. G., & Brogi, G. (2006). Fog detection using airport radar. *Proceedings of IRS 2006 International Radar Symposium,* pp. 209-212, *Krakow, Poland May 24-26, 2006,*
72. Gary Comparetto., 1993. The Impact of Dust and Foliage on Signal Attenuation in the Millimeter Wave Regime. *Journal of Space Comm*, vol. 11, no. 1, pp. 13-20.
73. George Edwards and Ravi Sankar , 2002. A Model for Analyzing Handoff in Cellular Communication Systems. *International Journal of Parallel and Distributed Systems and Networks,* vol. 5, no. 1, , pp. 1-6.
74. Ghassemzadeh S. S., R. Jana, C. W. Rice, W. Turin and V. Tarokh, 2011. A statistical path loss model for in-home UWB channels. *IEEE transactions on vehicular technology,* vol. 60, no. 1.
75. Ghosh P. M., Hossain Md. Anwar, Abadin A.F.M. Zainul,Karmakar K. K., 2012. Comparison Among Different Large Scale Path Loss Models for High Sites in Urban, Suburban and Rural Areas. International Journal of Soft Computing and Engineering (IJSCE), Volume-2, Issue-2, pp. 287-290.
76. Ghosh A., J. G. Andrews, R. Chen and D. R. Wolter, 2005. Broadband wireless access with WiMax/802.16: current performance benchmarks, and future potential. *IEEE Communications Magazine*, vol. 43, no. 2, pp. 129–136.
77. Gilhousen K.S, et. al., 1991. On the Capacity of a Cellular CDMA System. *IEEE Transaction on vt,* vol. 40, no. 2, pp.303-312.
78. Goldhirsh J., and R. L. Robinson,1982.Attenuation and space diversity statistics calculated from radar reflectivity data of rain. *IEEE Trans. Antennas Propagation.,* vol. AP-30.
79. Gold smith. Andrea, Wireless communications book, Cambridge university press.
80. Gordon L. Stüber,"Principles of Mobile Communication",Second Edition, Kluwer Academic Publishers, ISBN: 0-306-47315-1, pp. 4-10,2002.
81. Gorazd Kandus., Hrovat A., and Tomaž Javornik, 2012. Path Loss Analyses in

Tunnels and Underground Corridors. *International journal of communications.* Vol.6, issue3, pp.136-144.
82. Greg Durgin., and Theodore S. Rappaport., 1998. Measurements and Models for Radio Path Loss and Penetration Loss In and Around Homes and Trees at 5.85 GHz. *IEEE Transactions on communications,* vol.. 46, No. 11, pp.1484-1496.
83. Griffin Joshua D. and Durgin Gregory D., 2009. Complete Link Budgets for Backscatter-Radio and RFID Systems. IEEE Antennas and Propagation Magazine, Vol. 51, No.2, pp. 11-25.
84. E.V.D Glazier, H.R.L Lamont, "Transmission and propagation" Her majesty's stationary Office, vol 5, 1958.
85. Gupta V., Sharma S. C. and Bansal M. C., 2009. Fringe Area Path Loss Correction Factor for Wireless Communication. *International Journal of Recent Trends in Engineering,* vol. 1, no. 2, pp.215-217.
86. HarriHolma, AnttiToskala, "WCDMA FOR UMTS", Third Edition, John Wiley & Sons Ltd., ISBN: 0-470-87096-6,pp. 7-48, 2004.
87. Hata M, "Fourth Generation Mobile Communication Systems Beyond IMT-2000", in *Proceedings of the Fifth Asia-Pacific Conference on Communications (APCC'99),* 1999, pp. 765–768.
88. Haryana State Action Plan on Climate Change report, Government of Haryana, Dec 2011
89. Hata M, 1981. Empirical formula for propagation loss in land mobile radio services. *IEEE Transactions on Vehicular Technology,* vol. 29, pp. 317–325.
90. Haykins S. and Moher M., 2005. Modern Wireless Communication", Pearson Education, pp.105-125.
91. Hazer Inaltekin., Mung Chiang., H. Vincent Poor., and Stephen B. Wicker.,2009. On Unbounded Path-Loss Models: Effects of Singularity on Wireless Network Performance. *IEEE Journal on selected areas in communications,* vol. 27, No. 7, pp.1078-1092.
92. Helhel S., Ozen S. and Goksu H., 2008. Investigation of GSM signal variation dry and wet earth effects. Progress In Electromagnetics Research vol. 1(1), pp. 147- 157.
93. Hemani S. and M. Oussalah, 2006. Mobile Location System Using Net monitor and MapPoint server. *Proceedings of Sixth annual Post graduate Symposium on the Convergence of Telecommunication,* pp.17-22.
94. A. Hecker, M. Neuland, and T. Kuerner, "Propagation models for high sites in urban areas", Adv. Radio Sci., 4, pp. 345-349, 2006.
95. Hie Sng Sin, 2004. Radio Channel Modelling for Mobile AD HOC Wireless Networks. Master of Science thesis, Naval Postgraduate School, Monterey, California.
96. Hviid J. T., J. B. Andersen, J. Toftgard and J. Bojer, 1995. Terrain-based propagation model for rural areas - an integral equation approach. *IEEE Transactions On Antennas and Propagation,* vol. 43, no. 1, pp. 41-46.
97. Hyo-Sung Ahn., SangYoung Park., and Wonpil Yu., 2008. Adaptive Path-loss Model- based Indoor Localization. IEEE transactions (1-4244-1459-8/08/$25.00 ©2008 IEEE).

98. Ikegami F., S. Yoshida and M. Umehira, 1984. Propagation factors controlling mean field strength on urban streets. *IEEE Transaction on Antennas and Propagation,* vol. 32, no. 8, pp. 822-829.
99. Imranullah Khan, Tan Chon Eng, Shakeel Ahmed Kamboh, 2012.Performance Analysis of Various Path Loss Models for Wireless Network in Different Environments. *International Journal of Engineering and Advanced Technology (IJEAT)* ISSN: 2249 – 8958, Vol.2, Issue.1, pp.161-165.
100. IPS Radio and space services, Introduction to HF Radio Propagation, Australian government. PP. 1-25.
101. Isaac I. Kim ; Bruce McArthur and Eric J. Korevaar"Comparison of laser beam propagation at 785 nm and 1550 nm in fog and haze for optical wireless communications", *Proc. SPIE* 4214, Optical Wireless Communications III, 26 (February 6, 2001); doi:10.1117/12.417512; http://dx.doi.org/10.1117/12.417512
102. Iskander M.F. and Zhengqing Yun , 2002. Propagation prediction models for wireless communication systems. *IEEE Transactions on Microwave Theory and Techniques,* vol. 50, Issue 3, pp. 662 – 673.
103. Ishimaru Akira, 1978. Wave Propagation and Scattering in Random Media, vol.1, Acadamic Press, New York , 1978, pp. 41-68.
104. ITU-R Recommendation F.1490, 2000. Generic requirements for fixed wireless access systems. (http://www.itu.int/dms_pubrec/itu-r/rec/f/R-REC-F.1490-0-200005-S!!PDF-E.pdf).
105. ITU-R Recommendation SM.1708, 2005. *Field-strength measurements along a route with geographical coordinate registrations.*
106. ITU, "World Telecommunication Development Report 2002: Reinventing Telecoms", March, 2002, http://www.itu.int/itud/ict/publications/.
107. ITU-R Recommendation P.1546, 2001. Method for point-to-area predictions for terrestrial services in the frequency range 30 MHz to 3000 MHz. *International Telecommunication Union.*
108. ITU-R Recommendation P.838-3, Specific attenuation model for rain for use in prediction methods
109. ITU-R Recommendation, Attenuation due to clouds and fog, 1994, PN series
110. Ivanovs, G. and D. Serdega, "Rain intensity influence on to microwave line payback terms," *Electronics and Electrical Engineering*, No. 6(70), 60-64, 2006.
111. Jain A., Upadhyay R., Vyavahare P. D. and Arya L. D., 2007. Stochastic Modeling and Performance Evaluation of Fading Channel for Wireless Network Design. *IEEE International Conference AINA 07, PAEWN*, pp. 893-897.
112. J. Rajala, K. Siplilä, and K. Heiska, "Predicting in-building coverage for micro cells and small macro cells," in *Proc. IEEE Vehicular Technology Conf. (VTC99),* Houston, USA, May 1999, pp. 180–184.
113. Jiang Jonathan H. and Dong L. Wu, 2004. Ice and water permittivity for millimeter and sub-millimeter remote sensing applications. Atmospheric Science Letters, March 31, pp. 146-151.
114. Joe Montana (George mason University) Dr. James W. LaPean course notes Dr. Jeremy Allnutt course notes And some internet resources + Tim Pratt book.
115. John S. Seybold, 2005 .*Introduction to RF Propagation.* John Wiley & Sons, pp. 69-80.
116. Jordi Perez Romero, Oriol Sallent and Ramon Agusti , 2006. Enhanced Radio

Access Technology Selection Exploiting Path Loss Information. *IEEE 17th Annual International Symposium on Personal, Indoor and Mobile Radio Communications.*
117. Joseph Wout and Martens Luc, 2006. Performance evaluation of broadband fixed wireless system based on IEEE 802.16. *IEEE wireless communications and networking Conference,* Las Vegas, vol. 2, pp.978-983.
118. Josip Milanovic, Rimac-Drlje S. and Bejuk K, 2007. Comparison of propagation model accuracy for WiMAX on 3.5GHz. *14th IEEE International conference on electronic circuits and systems,* Morocco, pp. 111-114.
119. Jon. W. Mark, Weihua Zhuang, 2005, "Wireless Communications and Networking", Prentice-Hall, India.
120. Julio C. Costa., 2008. Analysis and optimization of empirical path loss models and shadowing effects for the Tampa Bay area in the 2.6 GHz band. Thesis report in Electrical Engineering Department of Electrical Engineering College of Engineering University of South Florida.
121. Karim M. R. and M. Sarraf, 2002. *W-CDMA and CDMA 2000 for 3G Mobile Network.* McGraw-Hill Telecom. Professional's pp. 332-334.
122. Keith T. Herring., Jack W. Holloway., David H. Staelin., and Daniel W. Bliss.,2010. Path-Loss Characteristics of UrbanWireless Channels. *IEEE Transactions on antennas and propagation,* vol. 58, No. 1, pp. 171-177.
123. Ken-Ichi, Itoh, Soichi Watanche, Jen-Shew Shih and Takuso safo , 2002. Performance of handoff Algorithm Based on Distance and RSS measurements. *IEEE Transactions on vehicular Technology,* vol. 57, no.6, pp.1460-1468.
124. Khairul Anuar Ishak, MATLAB Tutorial of Fundamental Programming.
125. Kishor S. Trivedi., Xiaomin Ma., and S. Dharmaraja., 2003. Performability modelling of wireless communication systems. *International journal of communication systems (*John Wiley & Sons) Ltd. 16:561–577 (DOI: 10.1002/dac.605).
126. Kunihiko Tsuboi and Nobutaka Okumura. A Next-Generation Workflow for System-Level Design of Mixed-Signal Integrated Circuits , Math works news letters. PP.1-4.
127. Kvicera Vaclav and Grabner Martin, 2010.Rain Attenuation on Terrestrial Wireless Links in the mm Frequency Bands, Advanced Microwave and Millimeter Wave Technologies Semiconductor Devices Circuits and Systems, ISBN: 978-953-307-031-5.
128. Laiho J., A. Wacker and T. Novosad (eds), 2006. *Radio network planning and optimization for UMTS , 2nd edition* , John Wiley & sons ,pp. 630.
129. Lakshmi Sutha Kumar, Yee Hui Lee, and Jin Teong Ong, "Truncated Gamma Drop Size Distribution Models for Rain Attenuation in Singapore"IEEE transactions on antennas and propagation, VOL. 58, NO. 4, April 2010, pp. 1325-1335
130. Liao D. and K. Sarabandi, 2007. Modeling and simulation of near-earth propagation in presence of a truncated vegetation layer. *IEEE Transactions on Antennas and Propagation,* Vol. 55, No. 3, pp.949-957.
131. Liebe H., 1981. Modelling the attenuation and phase of the radio waves in air at frequencies below 1000GHz" radio science vol;16 no 6 , pp 1183-1199.
132. Lkhagvatseren T, and Hruska F 2011. Pathloss aspects of a wireless communication system for sensors. International journal of computers and

communications, vol. 5, pp 18-26.
133. Lorne C. Liechty., 2007. Path loss measurements and model analysis of a 2.4 GHZ wireless network in an outdoor environment. Thesis report submitted at Georgia Institute of Technology. pp.1-52.
134. Löw K., 1992. Comparison of urban propagation models with CW-measurements. Proceeding *Vehicular Technology Conference VTC '92*, pp. 936-942.
135. Löw K., 1992. Comparison of CW-measurements performed in Darmstadt with the flat edge model. *COST 231 TD(92)8*, Vienna.
136. LUIGI MORENO, A Self-Learning E-Book based Corse, by redio engineering services, point to point radio link engineering.
137. Lukas Klozar, Jan Prokopec,"Propagation Path Loss Models for Mobile Communication", IEEE Proceedings of 21st International Conference Radioelektronika, ISSN: 978-1-61284-324-7/11, pp. 287-290, 2011.
138. Luo Lichum., 2000. A Mew MF and HF Ground-Wave Model for Urban Areas. *IEEE Antennas and Propagation Magazine,* Vol. 42, No. 1,pp. 22-33.
139. Maitham, Al-Safwani and Asrar U.H. Sheikh, 2003. Signal Strength Measurement at VHF. *the Journal for Science and Engineering,* vol. 28, no.2C, pp.3-18.
140. Maitra A., "Rain Attenuation Modeling From Measurements of Rain Drop Size Distribution in The Indian Region", *IEEE Antennas and Wireless Propagation Letters.* Vol. 3, P. 180–181, 2004.
141. Mardeni R. and Lee YihPey, 2010. The Optimization of Okumura's Model for Code Division Multiple Access (CDMA) System in Malaysia. *European Journal of Scientific Research,* vol.45, no.4, pp.508-528.
142. Mardeni.R., and Siva Priya.T., 2011. Performance of Path Loss Model in 4G Wimax Wireless Communication System in 2390 MHz. *International Conference on Computer Communication and Management IACSIT Press, Singapore.* vol.5, pp.546-550.
143. Mardeni R., and Lee Yih Pey., 2012. Path Loss Model Optimization for Urban Outdoor Coverage Using Code Division Multiple Access (CDMA) System at 822MHZ. *ISSN 1913-1844 E-ISSN 1913-1852* www.ccsenet.org/mas Modern Applied Science vol. 6, no. 1,pp.28-42.
144. Mardeni R. and K. F. Kwan, 2010. Optimization Of Hata Propagation Prediction Model In Suburban Area In Malaysia. *Progress In Electromagnetics Research*, Vol. 13, pp. 91-106.
145. Mawjoud S. A., 2008. Evaluation of Power Budget and Cell Coverage Range in Cellular Evaluation of Power Budget and Cell Coverage Range in Cellular GSM System. Al-Rafidain Engineering, Vol.16 ,No.1, 2008, pp.37-47.
146. Medeisis A. and A. Kajackas, 2007. Modification and tuning of the universal Okumura-Hata model for radio wave propagation predictions. *Asia-Pacific Microwave Conference 2007*, vol. 11, pp.1-4.
147. Meng, Y. S., Y. H. Lee and B. C. Ng, 2009. Study of propagation loss prediction in forest environment. *Progress In Electromagnetics Research*, vol. 17, pp. 117-133.
148. Meng Y. S., Y. H. Lee and B. C. Ng, "Further Study of Rainfall Effect on VHF Forested Radio-Wave Propagation With Four-Layered Model", Progress In Electromagnetics Research, PIER 99, 149-161, 2009.

149. Meng Yu Song, et. al., "The Effects of Tropical Weather on Radio-Wave Propagation Over Foliage Channel", *IEEE Transactions on Vehicular Technology*, vol. 58, no. 8, October 2009.
150. Meng Y. S.., and Y. H. Lee., 2010. Investigation of foliage effect on modern wireless communication systems: A Review. *Progress In Electromagnetics Research, vol. 105, pp.313-332.*
151. Meng Y. S., Lee Y. H., and Ng B. C., 2009. Empirical near ground path loss modeling in a forest at VHF and UHF bands. *IEEE Transactions on Antennas and Propagation,* vol. 57, no. 5,pp. 1461-1468.
152. Mohammed, S. et al., 2006. Measurements and Analysis for Signal Attenuation through Date Palm Trees at 2.1 GHz Frequency. Sudan Engineering Society Journal, Vol. 52, pp.17 -22.
153. Mohammed Alshami, Tughrul Arslan, John Thompson and Ahmet Erdogan., 2011. Evaluation of Path Loss Models at WiMAX Cell- edge. *IEEE Transactions* 978-1-4244-8704-2/11/$26.00 pp.1-5
154. Moinuddin A .A. and Singh S, 2007. Accurate Path Loss Prediction in Wireless Environment. *Institution of Engineers (India),* vol. 88, pp. 09 - 13.
155. Mukesh Kumar, Vijay Kumar, Suchika Malik., Performance and Analysis of Propagation Models for Predicting RSS for Efficient Hanndoff. *International Journal of Advanced Scientific and Technical Research*, Vol. 1 (2) (Feb. 2012) ISSN: 2249-9954, pp 61-70.
156. Nagendra Sah., Neelam Rup Prakash., and Deepak Bagai., 2012. Application CSP in Optimizing The Path Loss of Wireless Indoor Propagation Model. *International Journal of Engineering and Advanced Technology (IJEAT),* ISSN: 2249 – 8958, vol.1, Issue-4, pp. 91-95.
157. Nadir Z., Elfadhili N. and Touati F., 2008. Pathloss determination using Okumura- Hata model and spline interpolation for missing data for Oman. *In Proceedings of the World Conference on Engineering,* pp. 422 – 425.
158. Nadir Z., and M. Idrees Ahmed, 2010. Path loss Determination using Okumura-Hata Model and Cubic regression for missing Data for Oman. *International Conference on Communications Systems and Applications,* pp.804-807.
159. Nadir Zia., 2011. Seasonal pathloss modeling at 900 MHz for oman. *International Conference on Telecommunication Technology and Applications, Proc .of CSIT.* vol.5, pp187-191.
160. Nair R.A., 1993. Gain enhancement in dielectric core filled multimode conical horn antenna. *Antennas and Propagation Society International Symposium, 1993. AP-S. Digest* , vol.3, pp.1671 - 1674.
161. Naldongar, P., 2007. *An assessment of the Impact of Harmattan Particles on Microwave Propagation in the Savannah Region*, M.Sc thesis, Electrical Engineering Department, Ahmadu Bello University, Zaria.
162. Neyman, A.B "Study of short wave reception in Zaria" PhD dissertation, Electrical Engr Dept. A.B.U ,1981(pages 1-2; 62-63).
163. Ng'oma Anthony, 2005. Radio-over-Fibre Technology for Broadband Wireless Communication Systems. Ph. D. thesis, faculty of Electrical Engineering of the Eindhoven University of Technology.
164. Nobel D., May 1962. The history of land to mobile radio communications. *IEEE Transactions on Vehicular Technology*, pp. 1406-1416.
165. Noman Muhammad, Davide Chiavelli, David Soldani and Man Li, 2006. *QoS*

and QoE Management in UMTS Cellular Systems. John Wiley & Sons, Ltd. ISBN: 0-470-01639-6, pp. 1-8.
166. Noman Shabbir., Muhammad T. Sadiq., Hasnain Kashif., and Rizwan Ullah. 2011. Comparison of radio propagation model for long termevolution (LTE) netwok. *International Journal of Next- Generation Networks (IJNGN)*, vol.3, No.3, pp.28-42.
167. Ohmori S., 2000. The Future Generations of Mobile Communications Based on Broadband Access Technologies. IEEE Communications Magazine, pp.134 - 142.
168. Okumara Y., E.Ohmori, T.Kawano, and K.Faukudu, 1968. Field strength and its variability in VHF and land-mobile radio service. *Rev. Elec. Commun. Lab.*, vol.16, pp. 825-873.
169. Olagoke, O.T., 2005. Traffic Analysis of the Six Primary Centres in the Network of NITEL North/West Zone, Kaduna. MSc Thesis, Electrical Engineering Department, Ahmadu Bello University Zaria.
170. Oyesola Olayinka Olusola, "Seasonal variation of mobile radio propagation characteristics in KADUNA METROPOLIS AND ENVIRONS: A Case Study of MTN and AIRTEL" a thesis of Master of Science (M.Sc.) in Electrical Engineering in 2011.
171. Paier A., J. Karedal, N. Czink et al, 2007. Car-to-car radio channel measurements at 5 GHz: Path loss, power-delay profile and delay-Doppler spectrum. in *Proceeding International Symp. Wireless Communication System*, vol. 16, pp. 224–228.
172. *Pakistan Telecom Authority.* A report on Measurement of QoS For GSM International Roamer, 2007. Document no: 03-PTA-REP-QoS-01.
173. Parsons J. D., "The Mobile Radio Propagation Channel, Second Edition",John Wiley & Sons Ltd., ISBN: 0-470-84152-4, pp. 17-90, 2000.
174. Paul Brenner, Wren, " Tesla Against Marconi" ,*The Dispute for the Radio Patent Paternity.*www.teslasociety.com/pdf/tesla_against_marconi.pdf
175. Pavan Kumar V. S., Dr.B.Anuradha, Vivek, andNaresh., Improvement of Key Permormance Indicators and QoS Evalution in opernational GSM network. *International Journal of Engineering Research and Applications (IJERA)*, ISSN: 2248-9622 vol. 1, Issue 3, pp.411-417.
176. Pavlos F., Sofoklis K. and GEORGE K., 2007. Enhanced Handover Performance inCellular Systems based on Position Location of the Mobile Terminals. *Seminar paper, Telecommunications Laboratory,* National Technical University of Athens. pp.2-3.
177. Pollini g p, "Trends in Handover Design", IEEE Communications Magazine , pp. 82 – 90, March 1996 .
178. Popoola J. J.,I. O. Megbowon, and V. S. A. Adeloye3.,2009. Performance Evaluation and Improvement on Quality of Service of Global System for Mobile Communications in Nigeria. *Journal of Information Technology Impact,*vol. 9, No. 2, pp. 91-106.
179. Pozar David M.. Microwave Engineering, Second Edition. John Wiley & Sons, pp. 686-689.
180. Pu Wang., Zhi Sun., Mehmet C. Vuran., Mznah A. Al-Rodhaan., Abdulla M. Al-Dhelaan., and F. Akyildiz., 2011. On network connectivity of wireless sensor networks for sandstorm monitoring. Elsevier , journal of Computer Networks,

vol. 55, pp. 1150–1157.

181. Qing-An Zeng., and Dharma P., Agarwal., 2002. Handoff in Wireless Mobile Networks. *Handbook of Wireless Networks and Mobile Computing,* ISBN 0-471-41902-8 © 2002 John Wiley & Sons, Inc.
182. Rafiqul Islam MD., Zain Elabdin Elshaikh., Othman O. Khalifa., A.H.M. Zahirul Alam., and Sheroz Khan., 2010. Fade Margin Analysis Due to Duststorm Based on Visibility Data Measured in a Desert. American Journal of Applied Sciences, vol.7, pp. 551-555.
183. Ramage, C.S.: Monsoon Meteorology, 296 pp., Academic Press, New York 1971.
184. Ramjee Prasad, 1998. Universal Wireless Personal Communication. Artech House Mobile Communication Series.
185. Rao R. R., and Ephremides A., "On the stability of interacting queues in a multiple-access system," IEEE Transactions on Information Theory. Vol. 34, No. 5 (September 1988), pp. 918-930. *DOI: 10.1109/18.21216*
186. Rappaport T. S., 2005. *Wireless Communications: Principles and Practice.* 2nd edition, Prentice Hall, pp. 151-152 & 547-597.
187. Razak AB. ,Bin Mansor, 2005. Improved Path Loss Models for Radiowave Propagation in Suburban Environment in the Universiti Putra Malaysia Campus and Taman Sri Serdang. Master of Science thesis, School of Graduate Studies, Universiti Putra Malaysia.
188. Ray, P. S. (1972), Broadband Complex Refractive Indices of Ice and Water, *Appl. Opt.,* 11 (8), pp.1836-1844.
189. Reusens E., W. Joseph., G.Vermeen., L. Martens., and B. Latre., I. Moerman., B.Braem., and C.Blondia., 2007. Pathloss models for wireless communication channel along Arm and Torso: Measurements and Simulations. IEEE 1-4244-0878-4/020.00.
190. Ronald Henry Clarke and John Brown " Diffraction theory and Antennas" Ellis Horwood Limited, England, 1980.
191. Roelens L., 2005. Pathloss model for wireless narrowband communication near biological tissue. *Sixth FirW PhD symposium, faculty of eng., Ghent university,* paper no.120, pp.1-2.
192. Saleh I.M., and E.M. Abuhdima., 2011. Effect of Sand and Dust Storms on Microwave Propagation Signals in Southern Libya. *Journal of Energy and Power Engineering,* vol.5, pp.1199-1204.
193. Saunders S. R.,2007. *Antennas and Propagation for Wireless Communication System.* John Wiley & Sons, ISBN 978-0-470-84879-1, pp.110-125.
194. Saveeda P., E.Vinothini, Vardhi Swathi and K.Ayyappan, 2013. Received Signal Strength (RSS) Calculation for GSM Cellular System at BSNL Pondicherry using Modified HATA Model. International Journal of Science, Engineering and Technology Research (IJSETR) Volume 2, Issue 1, pp. 43-47.
195. Sebastian Büttrich, wire.less.dk edit: September 2009, Pokhara, Nepal, http://creativecommons.org/licenses/by-nc-sa/3.0/.
196. S.Sakagami,K,Kuboi, "Mobile Propagation Loss Prediction for Arbitrary Urban Environment",IEICE Trans.Commun.,Vol.J74-B-,No.1,pp.17-25,Jan.1991.
197. H.K.Sharma, S.Sahu, S.sharma, "Enhanced Cost231W.I.Propagation Model in Wireless Network" *International Journal of Computer Application (0075-8887) Volume 19-No.6, April 2011.*

198. Shittu, W.A. ,2006. Cellular Mobile Radio Propagation Characteristics: Case Study of Globacom and MTN. MSc Thesis. Electrical Engineering Department, Ahmadu Bello University, Zaria.
199. K. Siwiak, *Radiowave Propagation and Antennas forPersonal Communications*, Artech House, 1998, pp. 199-235.
200. Shoewu O. and Adedipe A., 2010. Investigation of radio waves propagation models in Nigerian rural and sub-urban areas. *American journal of scientific and industrial research,* vol. 12, pp. 227-232.
201. Shoewu, O. and Edeko, F.O.,2011. Outgoing call quality evaluation of GSM network services in Epe, Lagos State. *American Journal of Scientific and Industrial Research,* ISSN: 2153-649X doi:10.5251/ajsir.2011.2.3.409.417.
202. Shuler Alt, E. E. A simple expression for estimating attenuation by fog at millimetre wavelengths. *IEEE Transactions on Antennas and Propagation*, Vol.32, No.7, (July 1984), pp.757-758, ISSN 0018-926X
203. Simic I. lgor, Stanic I. and Zrnic B., 2001. Minimax LS Algorithm for Automatic Propagation Model Tuning. *Proceeding of the 9th Telecommunications Forum (TELFOR 2001.*
204. Smitha K.,Sivabalan , John J., *Int. J. Communications, Network and System Sciences*, 2009, 2, 754-758 doi:10.4236/ijcns.2009.28087 blished Online November 2009 (http://www.SciRP.org/journal/ijcns/). Copyright © 2009 SciRes. *IJCNS* Modified Ceiling Bounce Model for Computing Path Loss and Delay Spread in Indoor Optical Wireless Systems.
205. K. Siwiak, Radio Wave Propagation and Antennas for Personal Communications.Norwood, MA: Artech, 1995, pp. 141–144.
206. Sridhar M., K. Padma Raju and Ch. Srinivasa Rao, 2012. Estimation of Rain Attenuation based on ITU-R Model in Guntur (A.P), India. ACEEE International Journal on Communications, Vol. 03, No. 03 pp. 6-10.
207. Stefan krone., falko guderian., gerhard fettweis., markus petri., maxim piz., miroslav marinkovic., michael peter., robert felbecker., and wilhelm keusgen., 2011. Physical layer design, link budget analysis, and digital baseband implementation for 60 GHz short-range applications International Journal of Microwave and Wireless Technologies, pp. 1-12.
208. Sudhir Dixit, Yile Guo, and Zoe Antoniou, 2001. Resource Management and Quality of Service in Third-Generation Wireless Networks. *IEEE communication magazine* ,pp.125-133.
209. Tabhane Samil, 2000. Handbook of Mobile Radio Networks. Artech House Mobile Communication Library.
210. Tan I., W. Tang, K. Laberteaux and A. Bahai, 2008. Measurement and analysis of wireless channel impairments in DSRC vehicular communications. in *Proceeding IEEE International Conference on Communication,* pp. 4882–4888.
211. Tapan K Sarkar., Zhong Ji., Kyungjung Kim., Abdellatif Medouri., and Magdalena Salazar-Palma., 2003. A Survey of Various Propagation Models for Mobile communication. *IEEE Antennas and Propagation Magazine*, vol.45, No.3. pp.52-82.
212. Thomas L Frey, Jr., "The Effects of the Atmosphere and Weather on the Performance of a mm-Wave Communications Link", Applied Microwave & Wireless, Feb. 1999, pp. 76-80.
213. Thomsen J. and Manggard R., "Analysis of GSM Handover using coloured Petri

Net" Master thesis, university of Aarhus, Denmark in 2003.
214. Timothy, K. I.; Ong, J. T. &Choo, E. B. L. (2002), "Raindrop Size Distribution Using Method of Moments for Terrestrial and Satellite Communication Applications in Singapore", *IEEE Transactions on Antennas and Propagation*, Vol. 15, No. 10, October 2002, 1420- 1424, ISSN: 0018-926X
215. Tomar G.S and Verma S., 2006. Analysis of handoff initiation using different path loss models in mobile communication system. *Proceedings of IEEE International Conference on Wireless and Optical Communications Networks*, vol. 4.
216. Tranzeo wireless technologies Wireless Link Budget Analysis http://www.tranzeo.com/allowed/Tranzeo_Link_Budget_Whitepaper.pdf.
217. Trivedi D K, Naveen Kumar Chaudhary, , Roopam Gupta, " Radio Link Reliability in Indian Semi-Desert Terrain under Foggy Conditions" pubished in International Journal of Latest Trends in Computing (E-ISSN: 2045-5364) , Volume 2, Issue 1, March 2011, pp. 47-50
218. Turkan ERBAY DALKILIC., Berna Yesim HANCI., and Aysen APAYDIN., 2010. Fuzzy adaptive neural network approach to path loss prediction in urban areas at GSM-900 band. *Turk J Elec Eng & Comp Sci,* vol.18, No.6, pp.1077-1094.
219. Thiago, H., et al. 2001. Survey: system for propagation analysis in mobile communication environments, .
220. Ubom E.A., Idigo V. E., Azubogu, A.C.O., Ohaneme C.O. and Alumona T. L., 2011. Path loss Characterization of Wireless Propagation for South – South Region of Nigeria. International Journal of Computer Theory and Engineering, Vol. 3, No. 3, pp. 360-364.
221. Vakili V T & S.S. Moghaddam , "Optimum Selection Of Handoff Initiation Algorithm & Related Parameters", DSP Research Lab.
222. Veeravalli.V and Sendonaris, 1999. The Coverage–Capacity Tradeoff in Cellular CDMA Systems. IEEE *Transaction on VT*, vol.48, no.5, pp.1443-1450.
223. VenkataSai Sireesha B., Dr.S.Varadarajan, Vivek and Naresh., Increasing Of Call Success Rate In GSM Service Area Using RF Optimization . *International Journal of Engineering Research and Applications (IJERA),* ISSN: 2248-9622 Vol. 1, Issue 4, pp. 1479-1485.
224. Vinko Erceg, L. J. Greenstein, S. Tjandra, et al., 1999. An empirically based path loss model for wireless channels in suburban environments. *IEEE Journal on Selective Areas in Communication,* vol. 17, no. 7, pp. 1205–1211.
225. Vinko Erceg, K. V. S. Hari et al., 2001. Channel models for fixed wireless applications. technical report, *IEEE 802.16 Broadband Wireless Access Working Group.*
226. Vishal jindal, "TEMS Parameters" Space Teleservices Presentation.
227. Vivek Kamboj and D.K. Gupta, 2011. Comparison of Path Loss Models for WIMAX in Rural environment at 3.5 GHZ. International Journal of Engineering Science and Technology (IJEST), vol. 3, no. 2 , pp. 1432 – 1437.
228. Walfisch J.,and H. L. Bertoni, 1988. A theoretical model of UHF propagation in urban environments", *IEEE Transactions on Antenna s and Propagation,* vol. 36 (12), pp. 1788 – 1796.
229. Weinman J. A. and R. Davies, 1978. Thermal microwave radiances from horizontally finite clouds of hydrometeors. Journal of Geophysics Res., vol. 83,

pp. 3099-3107.
230. William C.Y. Lee, 1985. Estimate of local average power of a mobile radio signal. *IEEE Transaction on Vehicular Technology,* vol. 34, pp. 22-27.
231. William C.Y. Lee, 1995. *Mobile Cellular Telecommunications.* McGraw Hill International Editions.
232. William C.Y. Lee, 2008. *Mobile Communications Engineering-Theory and Applications*, Second edition. Tata Mc-Graw Hill Publishing company limited.
233. William keith, dishman, "Estimation of rain attenuation on earth space milli meter communications link" text book page no 46-51
234. Wu R. and J. A. Weinman, 1984. Microwave radiances from precipitating clouds containing aspherical ice, combined phase and liquid hydrometeors. Journal of Geophys. Res., vol. 89, pp. 7170-7178.
235. www.mathworks.com/products/matlab/.
236. X. Li, 2006. 'RSS-based Location Estimation with Unknown Pathloss Model. *IEEE Transactions on Wireless Communications,* vol. 5, issue 12, pp. 3626-3633.
237. Yu – Huei, T. Wen Shyang, H. and Ce Kuen S., 2009. The influence of propagation in a live GSM network. *Journal of electrical engineering,* vol. 7(1), pp. 1 – 7.
238. Zhi Ren., Guangyu Wang., Qianbin Chen., and Hongbin Li., 2010. Modelling and simulation of Rayleigh fading, path loss, and shadowing fading for wireless mobile networks. *ELSEVIER Z. Ren et al. / Simulation Modelling Practice and Theory 19 (2011),* pp. 626–637.
239. F. Zhu and J. McNair, "Optimizations for vertical handoff decision algorithms," in *in Proc. of IEEE Wireless Communications and Networking Conference (WCNC),* vol. 2, March 2004, pp. 867 – 872.
240. Zreikat A.I et al, 2004. A Comparative Capacity/Coverage Analysis for CDMA Cell in Different Propagation Environments. *Wireless Personal Communications,* vol. 28, pp. 205–231.

APPENDICES

APPENDIX-I

MAP OF INDIA & HARYANA WITH DIFFERENT CLIMATIC CONDITIONS

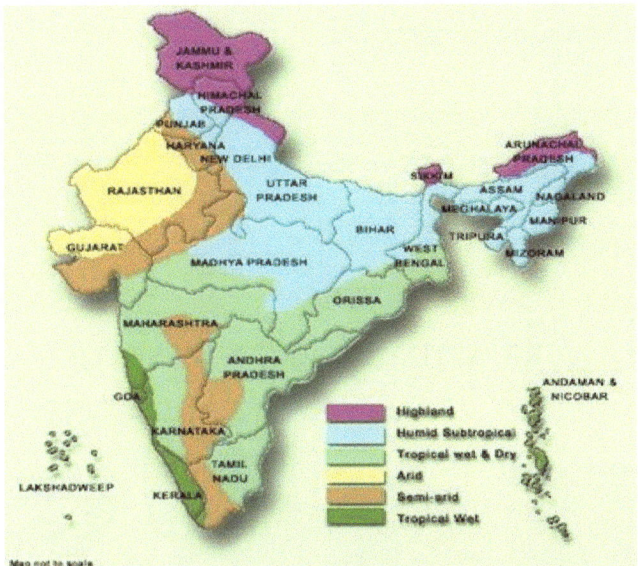

GEOGRAPHICAL MAP OF INDIA WITH DIFFERENT AREA

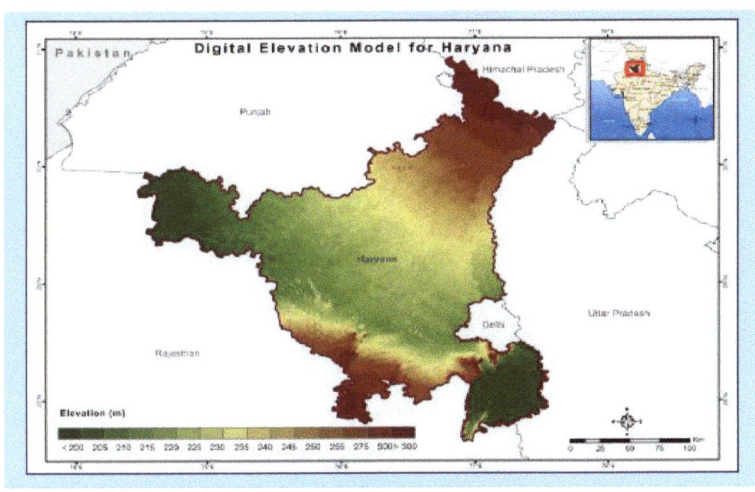

GEOGRAPHICAL MAP OF HARYANA

APPENDIX-II

DIFFERENT TERRIANS OF INDIA

RAJASTHAN TERRIAN (DESERT)

HILLY TERRIAN

SATELLITE VIEW & DIFFERENT CLIMATIC CONDITIONS OF NARNAUL (HARYANA TERRIAN)

APPENDIX-III

BTS SITES FOR FIELD DATA COLLECTION IN NARNAUL (HARYANA)

1. Base station (ID NNL001) - Lon. 76.1105 , Lat. 28.0413 ⎤
2. Base station (ID NNL002) - Lon. 76.1125 , Lat. 28.0427 ⎬ Urban Area
3. Base station (ID NNL003) - Lon. 76.1145 , Lat. 28.0445 ⎦
4. Base station (ID NNL004) - Lon.76.1154 , Lat. 28.0459   Sub urban Area
5. Base station (ID NNL005) - Lon. 76.1178, Lat. 28.0495   Rural Area

APPENDIX-IV

EXPERIMENTAL SETUP & TEMS NAVIGATION PERIPHERIALS

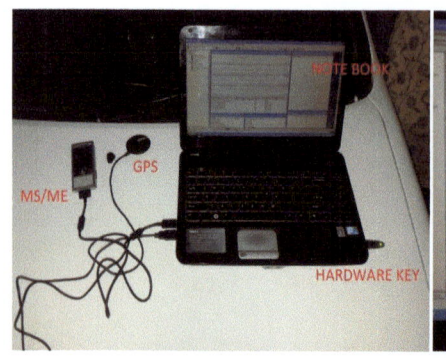
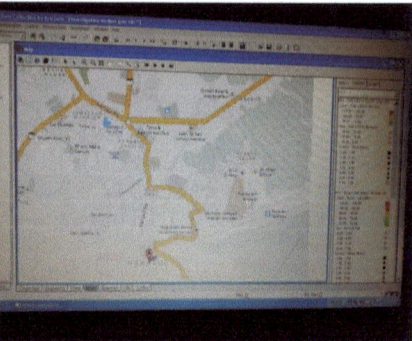

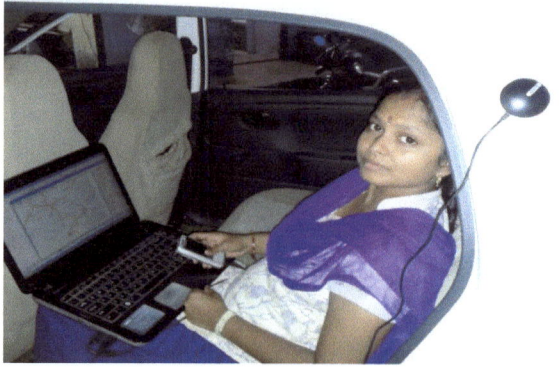

EXPERIMENTAL SETUP FOR DATA COLLECTION

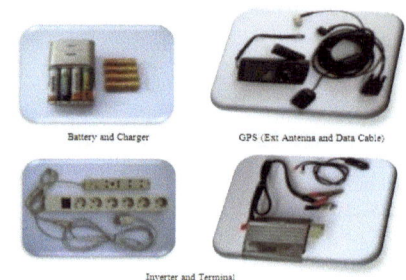
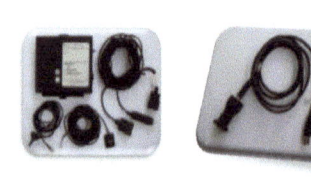

TEMS NAVIGATION PERIPHERIALS

APPENDIX-V

SIMULATION CHART

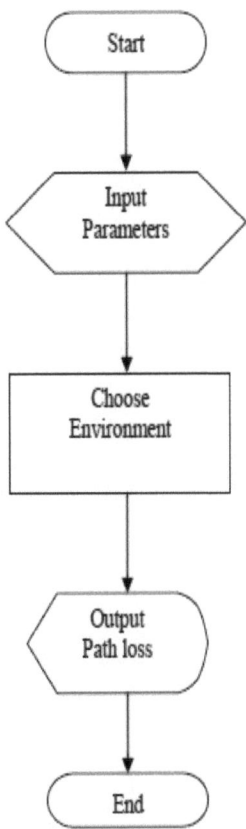

APPENDIX-VI

LINK BUDGET CALCULATIONS BY USING RADIO MOBILE SOFTWARE

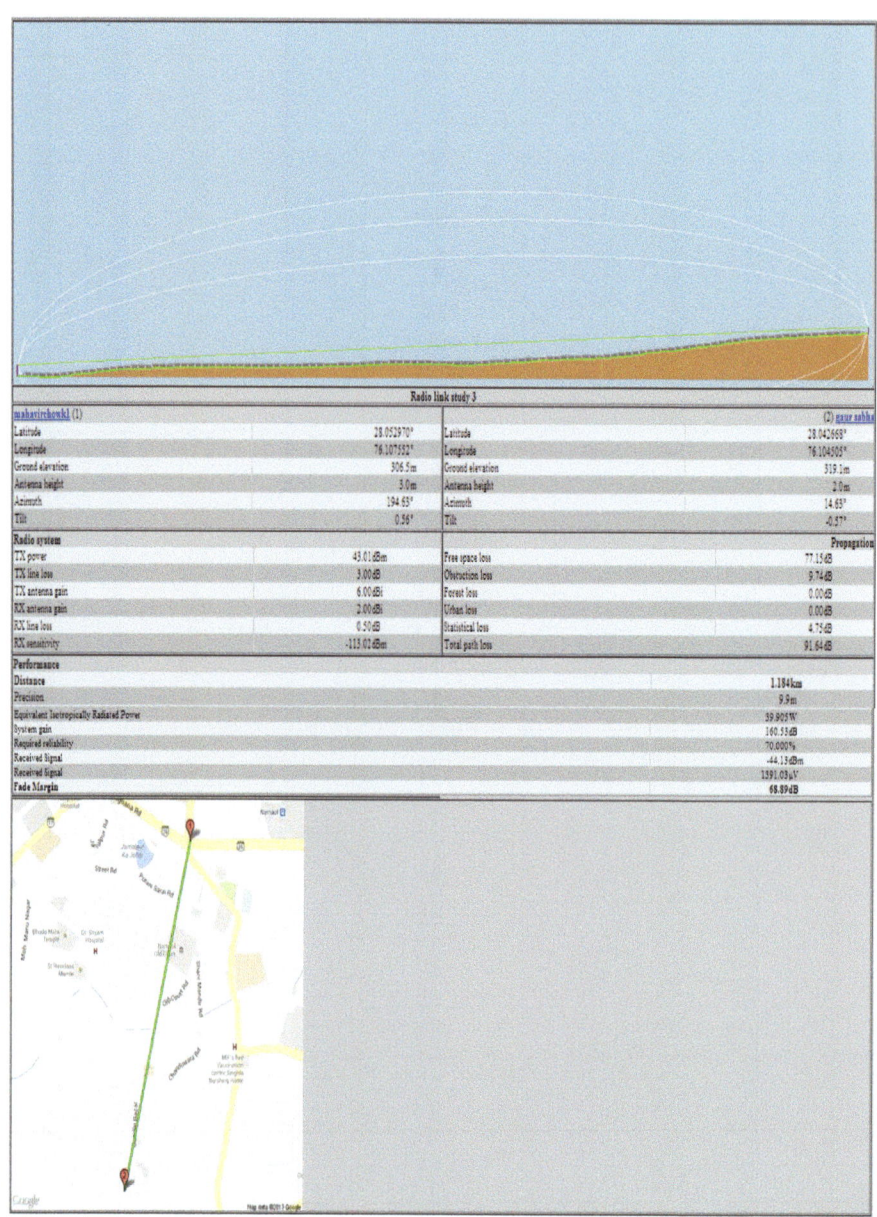